Blackness and Social Mobility in Brazil

"In this original and important contribution to the study of race, gender and class in Brazil, Doreen Gordon provides a rich ethnography of upwardly mobile Blacks in the northeastern city of Salvador. Despite the persistent scholarly denial that a Black middle-class exists, the author provides a necessary corrective to the idea that money whitens and lays the groundwork for unpacking the gender and class complexities of the Brazilian racial divide. This book illustrates how Black elites continue to be shaped by the intellectual and political work of Black social movements and cultural organizations precisely because racial violence structures their everyday and structural experiences. The author reminds us that even in a majority Black city and Black country, cultural dominance does not always translate into the eradication of racial inequality or guarantee full access to citizenship."

—Professor Keisha-Khan Perry, *The Presidential Penn Compact Professor of Africana Studies at the University of Pennsylvania, Philadelphia, Pennsylvania, United States*

"What are the practices, processes, and strategies that underlie black social mobility in contemporary Brazil? This book provides a broad and rich ethnography of Afro-Brazilian middle class families in the city of Salvador, northeastern Brazil. Informed by wide reading and detailed analysis, this work is relevant for all those interested in debates about race, inequality and social change."

—Professor Thais Machado-Borges, *Associate Professor in Latin American Studies, Stockholm University, Stockholm, Sweden*

Doreen Joy Gordon

Blackness and Social Mobility in Brazil

Contemporary Transformations

palgrave
macmillan

Doreen Joy Gordon
Department of Sociology, Psychology, & Social Work
University of the West Indies
Mona, Kingston, Jamaica

ISBN 978-3-030-90764-8 ISBN 978-3-030-90765-5 (eBook)
https://doi.org/10.1007/978-3-030-90765-5

This Palgrave Macmillan imprint is published by the registered company Springer Nature Switzerland AG.
The registered company address is: Gewerbestrasse 11, 6330 Cham, Switzerland

In memory of Derek, my father

Acknowledgements

This book is based on PhD fieldwork carried out in Salvador in the northeastern state of Bahia, Brazil. Anyone who has conducted research in a foreign country, particularly in a foreign language, knows how important it is to be able to turn to others, even regarding the simplest of experiences. This journey involved the patience, help, commitment, investment, and encouragement of many people, to whom I am forever indebted. First and foremost, I am especially grateful to the Brazilians and their families who kindly opened their homes and patiently answered my questions. Many led busy lives, yet sat with me through lengthy interviews and made efforts to include me in their activities. Some prayed for me and worried about my welfare in Brazil. I owe a debt of gratitude to these families and individuals that I will never be able to repay. I hope that I have represented them in all their complexities and take full responsibility for any shortcomings.

Second, I am especially grateful for the invaluable guidance provided by Peter Wade and John Gledhill at the University of Manchester: they were inspiring supervisors with whom to develop my ideas and often took a keen interest in my personal welfare. I would also like to thank Peter Fry and Angela Torresan for providing vital comments on my PhD thesis as well as the many colleagues and friends who helped along the way and had faith in my work. Dr Elena Calvo-Gonzalez and Dr Cecilia McCallum at the Federal University of Bahia provided invaluable practical and personal support. This research also benefitted from the collaboration and assistance of scholars based in Brazil such as Livio Sansone, Angela Figueiredo, Elisete da Silva, Jeferson Bacelar, Mariela Hita, and Luciana Duccini. I would like to express special thanks to Isabela Kikuta and Vera Rocha who provided research assistance in a professional, confidential, and caring manner. I thank Suzanna Gregory, Kiko Lisboa, Jacqueline Moreno, and Terrianna Selby for their language translation assistance at different stages of this project.

This research was made possible through the financial support of the University of Manchester, the Royal Anthropological Institute, and the University of the West Indies. Generous conference grants in the initial dissemination of this research were provided by the University of Manchester and the University of Oxford. My participation in the Human Economy programme as a Post-Doctoral Fellow at the University of Pretoria between 2010 and 2011 also helped me to think critically about my Brazilian research.

Institutional support was provided by the University of Manchester, where I completed my dissertation, as well as the University of the West Indies, Mona, where I have received grants and a Principal's Award to help with ongoing research and the writing of this book. I would also like to mention the assistance of the University of Texas at Austin libraries, especially the staff of the Nettie Lee Benson Latin American Collection. In addition, I would like to thank the Jamaican Embassy in Brazil for their help in facilitating interviews with senior civil servants on return visits to the country and the Brazilian Embassy in Jamaica for their ongoing support and advice. I also feel obliged to Palgrave Macmillan for offering me the opportunity to publish my first monograph, as well as to the anonymous reviewers at Palgrave Macmillan for their useful comments on my work.

Finally, I would also like to express my gratitude to close friends and family—especially my mother, Barbara, and sister, Haydee, who have provided me with patient and loving support and encouragement during these years. Special thanks to David Gauntlett, who supported my efforts by purchasing a camera for my fieldwork and paying for language lessons.

Summary of the Book

This book examines the emergence of black middle classes in urban Brazil, after 30 years of black mobilization and against the backdrop of deep economic, cultural, and political transformations taking place in recent decades within the country. One of the consequences of such transformations is said to be the re-structuring of gender, race, and class relations. Utilizing qualitative techniques such as ethnography, interviews, life histories, and focus groups in the Northeast region of the country, the book explores contemporary race, class, and gender inequalities and their impact on daily lived experience. It reveals the social dynamics underlying upward mobility, the diverse modes and experiences of social ascent into the middle classes, and the everyday negotiations involved in establishing one's status in Salvador's socio-racial hierarchy which are not captured by other, more "macro" lenses. While some of these patterns and experiences are not peculiar to black people, this book argues that race shaped the contours and possibilities of social mobility in particular ways. This book is critical reading for specialists in the fields of inequality, race, gender, and class relations.

Contents

List of Figures

List of Maps

List of Tables

Prologue

We are tied together in the single garment of destiny, caught in an inescapable network of mutuality. And whatever affects one directly affects all indirectly. (Martin Luther King Jr, from the speech "*Remaining Awake Through the Great Revolution*," 1968)

Following the World Conference against Racism, Racial Discrimination, Xenophobia and Related Intolerance held in 2001 in Durban, South Africa, the United Nations declared 2015–2024 to be the International Decade of the Peoples of African Descent with the theme, "People of African Descent: Recognition, Justice and Development." Indeed, it is an appropriate time to reflect on the ease with which race and racial discrimination has raised its head across the globe in recent times. We have witnessed mass protests across the United States and the rest of the world in support of the Black Lives Matter movement, especially following the death of George Floyd, an African American who was killed in police custody in Minneapolis on May 25, 2020. Despite an unprecedented global pandemic that has reshaped our lives as we know it, people in different parts of the world have filled the streets, government buildings, parks, and social media platforms calling for an end to systemic racism and sparking heated debates about histories of colonialism and slavery, the meanings of statues and monuments, acts of racism and police brutality, racial inequality, reparatory justice, and police reform. In Brazil, the protests have been dubbed "Vidas Negras Importam," highlighting the violence meted out to largely poor, black communities by the state and security forces. Meanwhile, an unstable political and social situation unfolded in the

United States, as the world watched the violent storming of the Capitol Building in Washington DC on January 6, 2021, by supporters of President Trump. In more recent months, social protests have spread across major cities and towns in Colombia as citizens (especially young people) make legitimate claims relating to social inequalities, structural racism, and exclusion—conditions which have been exacerbated by the pandemic.[1] However, these protests have been met with state repression and the undemocratic blocking of the entry into Colombia of international commissions and observers from the Inter-American Commission on Human Rights.[2] As Angela Davis, the political activist, philosopher, professor, and author, recently stated, this is an extraordinary moment for tackling racial injustice.[3] What are the lessons to be learnt from historical tragedies in global history that are/were related to racial and ethnic hierarchizing and discrimination (such as the Holocaust, the Atlantic Slave Trade, Apartheid, and the Rwandan Genocide)? How can these lessons be applied to anti-racist organizing today?

The global community has made some strides in tackling racism and racial discrimination. This includes the observance, on March 21, of International Day for the Elimination of Racial Discrimination, to commemorate March 21, 1960, when police opened fire and killed 69 people at a peaceful demonstration in Sharpeville, South Africa, against Apartheid pass laws. Other global efforts include the Universal Declaration of Human Rights (1948); the United Nations Declaration on the Rights of Indigenous Peoples (2007); and three world conferences to combat racism and racial discrimination held in Geneva in 1978 and 1983 and Durban in 2001. Yet continued vigilance is required. Recently, a global

[1] BBC News, "Why Colombia's Protests are unlikely to fizzle out." May 31, 2021. https://www.bbc.com/news/world-latin-america-56986821. Accessed June 14, 2021.

[2] Organization of American States, "IACHR requests authorization to conduct a Working Visit to Colombia in the Wake of Alleged Human Rights Violations During Social Protests." Organization of American States Media Center, May 14, 2021. http://www.oas.org/en/IACHR/jsForm/?File=/en/iachr/media_center/PReleases/2021/125.asp. Accessed June 14, 2021.

Borda, Sandra. "Why Colombia has erupted in Protest." *Open Democracy*. https://www.opendemocracy.net/en/democraciaabierta/colombia-protest-why-erupted-en/. Accessed June 14, 2021.

[3] Angela Davis quote cited in an article by Tonya Morley and Allison Hagan writing in "*Here and Now*," a news publication of Boston's NPR news station, WBUR (public radio). https://www.wbur.org/hereandnow/2020/06/19/angela-davis-protests-anti-racism. Accessed June 16, 2021.

call was made for concrete action for the elimination of racism, racial discrimination, xenophobia, and related intolerance as well as the comprehensive implementation of and follow up to the Durban Declaration and Programme of Action. While the resolution was adopted by the UN General Assembly on December 31, 2020, the vote summary is of some concern. Jamaica, Barbados, Brazil, Belize, Mexico, and Rwanda voted "yes"—however, powerful and influential countries such as Canada, the Netherlands, Israel, the United Kingdom, and the United States voted "no." Sweden, New Zealand, Nigeria, Norway, and Portugal abstained from voting.[4]

More recently, the United Nations Human Rights Council has urged global action including reparations to "make amends" for racism against people of African descent. This is based on a study carried out by the United Nations in the wake of George Floyd's death. In June 2020, the UN Human Rights Council adopted unanimously a resolution brought by African nations that condemned discrimination and violent policing and requested a report on systemic racism.[5] Based on discussions with more than 300 experts and people of African descent, nations were strongly encouraged to take action to end racial injustice. The report sites concern related to racial injustice in about sixty countries including the United Kingdom, Belgium, France, Brazil, and Colombia. Furthermore, the use of racial profiling and excessive police force was found to be systemic in much of North America, Europe, and Latin America. The report concludes that in order to achieve racial justice, countries should make amends for centuries of violence and discrimination, including through formal acknowledgement and apologies, truth-telling processes, educational reforms, memorialization, and reparations in various forms.[6]

[4] See United Nations Digital Library: A Global Call for Concrete Action for the Elimination of Racism, Racial Discrimination, Xenophobia, and Related Intolerance. https://digitallibrary.un.org/record/3896183?ln=en. Accessed June 16, 2021.

[5] Sammy Westfall, "UN Human Rights Chief Calls for Reparations to Address Systemic Racism around the World." *The Washington Post*, June 28, 2021. https://www.washingtonpost.com/world/2021/06/28/united-nations-systemic-racism-report/. Accessed July 30, 2021.

[6] BBC News, "UN Human Rights Chief Calls for Reparations Over Racism." June 28, 2021. https://www.bbc.com/news/world-europe-57646933. Accessed July 30, 2021.

The Situation of African Descendants in the Americas

According to the United Nations, around 200 million African descendants are living in the Americas, whether as descendants of former slaves or more as recent migrants.[7] They account for a quarter of Latin America's population and are over-represented among the poor, as highlighted by a new World Bank report.[8] Brazil has the largest African-descendant population in the world outside of Africa—indeed, roughly half of the Brazilian population identify as *preto* (black) or *pardo* (brown/mixed race), corresponding to about 107 million people out of a national population of 201 million (PNAD 2013).[9] Despite making up the majority of the population, however, statistics show that Afro-Brazilians are twice as likely to be poor as white Brazilians, have less access to quality health care and education, and are disproportionately affected by crime and violence.[10] Afro-Brazilians also appear to have fewer chances for social mobility—for example, a study carried out across major Brazilian cities and released in 2013 by the Inter-Union Department of Statistics and Socioeconomic Studies (DIEESE) indicated that black professionals earn 36 per cent less than their non-black counterparts—regardless of region or educational attainment.[11] Such statistics are bringing awareness to Brazil's systemic racism, long embedded as normal practice within Brazil's institutions, structures, and organizations.

While researchers have increasingly documented how racism affects the lives of African descendants in the Americas, there has been a long history of political and social elites promoting policies of "whitening" of the

[7] United Nations International Decade for the People of African Descent 2015–2024. https://www.un.org/en/events/africandescentdecade/index.shtml. Accessed July 30, 2021.

[8] World Bank Report on Afro-Descendants in Latin America (2018). http://documents1.worldbank.org/curated/en/896461533724334115/pdf/129298-7-8-2018-17-29-37-AfrodescendantsinLatinAmerica.pdf. Accessed July 30, 2021.

[9] *Pesquisa Nacional por Amostra de Domicilios* (2013), Instituto Brasileiro de Geografia e Estatística. https://biblioteca.ibge.gov.br/visualizacao/livros/liv94414.pdf. Accessed September 27, 2020.

[10] *World Bank Report on Afro-Descendants in Latin America* (2018). http://documents1.worldbank.org/curated/en/896461533724334115/pdf/129298-7-8-2018-17-29-37-AfrodescendantsinLatinAmerica.pdf. Accessed September 27, 2020.

[11] *Os Negros nos Mercados de Trabalho Metropolitanos*. 2013. Departamento Intersindical de Estatística e Estudos Socioeconômicos. https://www.dieese.org.br/analiseped/2013/2013pednegrosmet.pdf. Accessed September 27, 2020.

population,[12] downplaying the contributions of Afro-Latin Americans, and denying the significance of racism—effectively rendering black voices and bodies invisible (Andrews, 2004; Perez-Sarduy and Stubbs, 1995; Wade, 2010). Traditionally, Latin Americans have tended to see themselves as societies living in different versions of "racial democracies," defined by the harmonious blending of races and cultures (Wade, 2010; Wade, Scorer and Aguiló, 2019). The advent of the racial democracy myth in Brazilian society in the early twentieth century allowed successive governments to brush off racism as an outdated problem that had been "solved" by slavery's abolition. The result is a society where racism and resulting social inequalities are not revealed or discussed, so it seems not to exist. Indeed, the black activist Edna Roland, who was a Rapporteur at the World Conference against Racism in Durban in 2001, likened racism in Brazil to a "Hydra of Lerna," a mythological, multi-headed creature that when you cut one of its heads off, other heads appear immediately in many places and positions (Ciconello, 2008: 1). The result is that racism becomes embedded in various institutional structures and patterned social relations, developing in new and different forms.

In more recent years, there has been a sea change in attitudes and practices regarding race in Brazil. Since Brazil's democratic opening in the 1980s, there have been determined efforts by black activists, researchers, and academics to bring attention to persistent racial inequalities in Brazil, along with increased public awareness and political attention to the issue. For example, the government implemented a mix of policies meant to promote social inclusion such as affirmative action, reserved quotas in the job market and in educational institutions, awareness campaigns, and anti-discrimination legislation. In particular, the creation of the Palmares Foundation—a federal institution named after a well-known *quilombo* settlement in colonial Brazil—was widely regarded as a victory for the

[12] Rooted in different strands of scientific racism and eugenics thinking at the time, late nineteenth- and early twentieth-century elites in Latin America promoted policies of "whitening" for their obviously non-white populations. This included encouraging European emigration to Brazil and ascribing a higher value to European cultural practices amongst others. Their hope was that black Brazilians would have children with white Europeans and their descendants, producing generations of lighter-skinned Brazilians. This idea of "whitening" was understood to be positively aligned with the aim of creating a modern, civilized Brazil, thereby eliminating the threat of degeneration and backwardness which were thought to be the result of a high degree of racial mixture in the population.

Afro-Brazilian community.[13] The institution represented a powerful symbol of the Brazilian state's recognition of the consequences of slavery and its commitment to work towards racial equality, while recognizing and protecting the Afro-Brazilian perspective as a key element in the country's history. This recognition of the Afro-Brazilian contribution to Brazilian history and society has improved the general awareness in the population of racism and racial inequalities in Brazil (Ciconello, 2008).

However, the gains that Brazil has made in addressing racial inequalities since at least the 1980s have been increasingly threatened by current events and politics in Brazil, including a growing opposition among some sectors of the population to legislative changes and redistributive policies—such as affirmative action—aimed at improving and protecting the rights of Afro-Brazilians.[14] The resurgence of Far Right politics in Brazil, intensifying with the election of Jair Bolsonaro in October 2018, is being propelled by a wave of class resentment, racism, sexism, homophobia, and religious fundamentalism, in reaction to the long-standing dominance of the Workers' Party (2003–2016) and their public policies (Perry 2020; Vargas 2016). One of the more troubling developments in recent times is that the man tasked with promoting black culture in Brazil, Sérgio Camargo, has made several inflammatory remarks, such as calling the Brazilian Black Movement "scum" and threatening to cut funding for the annual Black Consciousness Day events.[15] Additionally, historically important academic research centres dedicated to studying Afro-Brazilian history and culture and Brazil-Africa relations are now under increased pressure due to shifts in foreign policy, lack of funding, resources, and full-time staff, among other challenges.[16] Recent developments like these

[13] A *quilombo* is a settlement that was formed by enslaved people and others escaping bondage and oppression in colonial Brazil. The name of this particular maroon settlement was called Quilombo dos Palmares, which was located in the Northeast of Brazil in the modern-day state of Alagoas.

[14] See Rafael Lima, "How Brazil's Anti-Racism Agency became a part of the Problem" in *The Brazilian Report* published on July 6, 2020. https://brazilian.report/society/2020/07/06/brazil-anti-racism-agency-became-part-of-the-problem/?fbclid=IwAR2bv4INGsUMA-RK-DZRE4tMdMltuys0_M6eqgSojWo288_iKNh9xjZ2qMM. Accessed July 9, 2020.

[15] Tom Phillips, "Black Official recorded calling Black Rights Movement Scum," *The Guardian*, June 3, 2020. https://www.theguardian.com/world/2020/jun/03/brazil-sergio-camargo-black-rights-movement-scum-recording. Accessed September 27, 2020.

[16] Two historically important centres of African research in Brazil are the Center for Afro-Oriental Studies in Salvador and the Center for Afro-Asian Studies in Rio de Janeiro. The former was founded in 1959 at the Federal University of Bahia. The latter was originally

make it all the more urgent to understand contemporary Brazilian racial dynamics in the effort to dismantle racism.

What This Book Is About

This book draws on long-term ethnographic research and engagement with a specific location—Salvador, Bahia, Brazil. At the same time, it links with larger processes occurring across the African diaspora, suggesting that the world can learn from Brazil's long history of the denial of racism and its efforts to dismantle it. This book therefore defends the possibility of discussing race and racism within a diasporic continuum, attuned to commonalities and distinctions across the black diaspora.

Many studies on race relations in Brazil have focused on marginalized and excluded populations, even as it is recognized that Afro-descendants constitute a heterogeneous population and their situations are not the same everywhere. Indeed, across different historical periods and regions in Latin America, there have been segments of the African-descendant population that have held higher social positions, despite enormous societal challenges and the pervasive presence of racism. Therefore, this book aims to bring fresh insight into the issue of black social mobility in contemporary urban Brazil, focusing in detail on the lives of a small number of upwardly mobile Afro-Brazilians who self-identified as *negro* (black) living in Salvador, Bahia. It fills an important need in the English language literature on black social mobility in Brazil for detailed, ethnographic analysis of the social backgrounds, economic strategies, social networks, daily lives, discourses, and practices of contemporary middle-class Afro-Brazilians—a group paradoxically marked by economic progress and social marginality. It is based on ethnographic fieldwork carried out between 2005 and 2007, with follow-up research in the summers of 2013 and 2017. It is important to note that this research has since been disseminated and published into various articles and book chapters, and that some of this material is reflected here.

Moreover, this book revises key narratives that have dominated public discourse in the past, such as the idea that African descendants "whiten"

established at Itamaraty under the Jânio Quadros Presidency and later moved to Cândido Mendes University. Both centres represented a desire for greater exchanges between Brazil and African countries, especially under the Lula Presidency. However, recent shifts in foreign policy have seriously undermined these efforts.

as they become upwardly mobile, discarding their links with their communities of origin and marrying into white families or adopting the lifestyles and values of the Euro-Brazilian elite. Using the concepts of "whitening" and "blackening"—as both symbolic and materialistic processes—I argue that contemporary black elites appear to be emerging in ways that challenge the traditional discourse, against the backdrop of a country that has experienced roughly 30 years of black mobilization and the introduction of neoliberal and multicultural policies across many Latin American and Caribbean nations. These kinds of processes indicate the need for a more qualitative view of black social mobility as lived experience, and the social consequences of such lived experiences. This could include research participants reacting to racial discrimination or stigmatization by staking their own claims to respectability and social value which may or may not be articulated through racialized discourses and identities.

In addition, I discuss the specific mechanisms and strategies of social ascension and the factors affecting the life choices, possibilities, and trajectories of research participants. The focus of studies on race and social mobility in Brazil has tended to be on individuals, leaving various questions and angles relatively unexplored that my research sought to address by focusing on the social contexts in which my "informants" live. My ethnographic material sheds light on the social processes, relationships, and networks underlying black social mobility, thereby nuancing dominant narratives about their integration into Brazilian society. The questions that engaged me while carrying out this research included: what are the actual practices and processes that underlie black social mobility in Brazil? Is it shaped by racism today (or in the past)? What modes of mobility are employed and are these changing? What are the opportunities and challenges encountered as persons confronted unequal society in Brazil? By focusing on upwardly mobile Brazilians who self-identified as *negro* (black), this book aims at providing a fresh angle on how people create, cope with, negotiate, or resist unequal society on an everyday basis. More broadly, it also throws light on the question of non-white elites in emerging economies such as Brazil and their ***potential*** to challenge traditional structures of power and inequality on a global scale.

Conducting Diasporic Research

The increased entrance of women, people of colour, LGBT and Third World scholars into the social sciences and humanities has enlivened debates about the fieldwork encounter and raised new existential and epistemological questions.[17] Building on this work, I will briefly outline my approach and positionality in relation to the research presented in this book. I was born into an educated middle-class family during the politically turbulent 1970s in Kingston, Jamaica, where discussions on inequality and social justice constantly took place. This personal context as well as my readings of a diverse collection of literature on African and African-diaspora communities shaped my interests in themes of race, class, and inequality in different post-colonial contexts and across language barriers. Brazil was of interest to me because of its large population of African descent, its history of slavery, its persistent racial and economic inequality, and its regional influence and power. Furthermore, I was inspired at the time by Brazil's political commitment to reduce social inequalities through programmes such as the "Fome Zero" (Zero Hunger) and "Bolsa Familia" (Family Stipend). I decided to carry out my PhD dissertation fieldwork in the north-eastern city of Salvador, due to its significant African-descendant population and its legacy of research on race relations.[18]

A Post-Doctoral Fellowship in 2011 at the University of Pretoria, South Africa, added a transnational dimension to my studies on race. For a year and a half, I participated in workshops, lectures, and conferences and studied the rise of black elites after Apartheid, which has continued to be the subject of heated debate in that country. One of the more memorable experiences I had in South Africa was visiting the Freedom Charter memorial in Walter Sisulu Square, based on a document ratified at the Congress of the People held in Kliptown, Soweto, in June 1955. The square has now been declared a heritage site and features an open-air museum that explains how the Freedom Charter—essentially a programme

[17] To have an example of this kind of work, see Anzaldúa (1981, 1990), Caldwell (2007), Diversi and Moreira (2009), hooks (1981, 1990, 1992), Hordge-Freeman (2015), Narayan (1993), and Twine and Warren (2000).

[18] The rich legacy of research on race relations in Salvador includes studies such as Azevedo (1996 [1953]), Figueiredo (2003), Frazier (1942), Guimarães (1987), Herskovits (1943), Landes (1947), Pierson (1942), Sansone (2003), Santana (2009), and Wagley (1952).

of non-racial social democracy—was written.[19] It also includes a monument—a conical brick tower at the centre of the square—that showcases the principles of the Freedom Charter, such as a demand for a multiracial, democratically elected government; equal opportunities; the nationalization of banks, mines, and heavy industries; and the redistribution of land.

While the Freedom Charter process was not perfect (e.g., Africanist members of the African National Congress rejected it and broke away to form the Pan Africanist Congress), it was an impressive attempt at capturing the demands of the people and articulating an alternative vision of South Africa. The process entailed collecting, collating, and representing the voices of ordinary South Africans and using it as a broad framework to inform policy. Its continued significance is that it holds global lessons for anti-racist and social justice organizing, as it represents an attempt to incorporate a more inclusive, bottom-up approach and active citizenship as the basis for addressing the challenges, needs, and aspirations of different groups of people. The post-Apartheid state made remarkable strides in adopting non-racialism, negotiating reforms to make the political system more inclusive, and encouraging participation in Truth and Reconciliation meetings. Nevertheless, high levels of inequality and xenophobic violence have plagued the new South Africa—so there is more work to be done.

I recognize that my experience as a female scholar from the Third World self-identifying as black means that I produce knowledge that is situated (Haraway 1988).[20] My understanding of the term "African diaspora" is that it is "a space of mutual recognition of a solidary consciousness across fragmented geographies" (Rahier et al. 2010, xviii). Indeed, the African diaspora has become a fundamental way of conceptualizing the experience of African-descendant communities, in both an academic sense and a political sense. This book attempts to make a contribution to critical and comparative race studies by engaging with diaspora theories, critical race theory, black feminist thought, and colonial and post-colonial perspectives to bring insights about Afro-Brazilian agency—particularly in terms of understandings of race, racialized identities, articulations of anti-racist politics, and gendered-class consciousness.

[19] The process was organized by the Freedom Alliance, comprising the National Action Council of the African National Congress, South African Indian Congress, South African Coloured People's Organization, and the South African Congress of Democrats.

[20] I recognize that in Brazil and Jamaica I would likely be identified by others as brown, mixed race, or white in some contexts. However my personal experiences and socialization has shaped my racial identification as black.

While there are differences across and even within national contexts, similar conditions and experiences affect black diasporic populations across the region and are entangled in global and national structures of inequalities and the circulation of ideas about "race" and difference. Africa and the African diaspora have tended to be ranked at the bottom of global hierarchies of races, ethnicities, cultures, and nations (Basch, Glick Schiller and Szanton-Blanc 1994; Thomas and Clarke 2013). This global racial-cultural hierarchy places Anglo-American culture at the apex and Sub-Saharan African culture at the base. Other cultures (usually with their own complex and dynamic internal hierarchies) jostle to occupy intermediate positions between the two extremes (Brodkin 1999; Frankenberg 1997). Global discourses about races and cultures which perpetuate the superiority of whiteness and the inferiority of blackness clearly affect the contours of local hierarchies—even while emerging non-white elites on the global scene potentially challenge Western hegemony.[21] This should lead us to think more critically about the centrality of racial negotiations for African descendants around the globe and the need to initiate more sustained dialogues across diasporic communities about how racialized systems are maintained, negotiated, accepted, and resisted. This is not to negate the specifics of Brazilian social relations—rather it is to analyse them in the context of an over-riding diasporic anti-blackness that structures social worlds in related ways. My approach to diasporic research and how it shaped my positionality in the field is dealt with in more detail in Appendix A of this book.

A Note on Terminology

Some basic terminological clarifications will help in the reading of this book. Firstly, I use italics to indicate terms or phrases in Portuguese that are used in everyday language, except for the names of people, places, or events in Brazil. Secondly, the study of African descendants in Latin America faces many challenges, beginning with the lack of agreement on who is and who is not Afro-descendant or black across and even within countries. Some scholars have commented that to talk about blacks as an

[21] By "emerging elites," I refer to the literature on the rise of the middle classes said to be taking place in the developing world, particularly in the emerging economies of Brazil, South Africa, India, and China (Davis 2010; Ncube and Lufumpa 2014; Neri 2008; Zhang 2010). A recent World Bank report (Ferreira et al. 2013) stated that Latin America and the Caribbean registered a 50 per cent jump in the number of people joining the middle classes in the last decade. More specifically, the middle classes in the region grew to an estimated 152 million in 2009, compared to 103 million in 2003.

ethnic group, where race is the basis for social identities, is a recent phenomenon (Guimarães 2013). Certainly, at different periods in Brazilian history, racialized identities emerged as significant—such as during the abolitionist movement, the black protests of the 1930s, or redemocratization movements after World War II and during the 1970s. Various names were used—such as *pessoas de cor* (people of colour), *pretos* (blacks), or *negros* (blacks), revealing different political projects (Guimarães 2013, 1). Thus, racial terminologies carry politicized meanings and can be challenging to navigate for the researcher of race relations. In this book, I adopt the current meaning deployed by Brazilian black activists and scholars that merges the *preto* (black) and *pardo* (brown/mixed race) colour/race categories on the national census into one category, *negro* (black).

When I am referring to the "Brazilian Black Movement" or "Black Lives Matter Movement," I have used uppercase letters to stress the political character of the movement. "African descendant" is a term that has generally been adopted by regional organizations in the wake of the Durban Conference. It describes people united by a common ancestry but living in dissimilar conditions, from Afro-indigenous communities such as the Garifuna in Central America to large segments of mainstream society, such as *pardos* (brown, mixed-race people) in Brazil. However, the term is rarely used in everyday discourse. Rather, words associated with being black or dark-skinned, like *negro* and *preto*, or terms referencing a mixed-race identity—such as *moreno*, *pardo*, *mestiço*, *mulato*, and *creole*—are much closer to Latin Americans' understanding of race relations.

In the Brazilian context, data on race is collected by the national census using five race/colour categories—*branco* (white), *pardo* (brown and/or mixed race), *preto* (dark-black), *amarelo* (yellow), and *indígena* (indigenous).[22] However, researchers have indicated that everyday racial classification is more flexible, ambiguous, and situational. Brazilian folk concepts of race are said to involve assessing and categorizing individuals according to characteristics such as colour of the skin, hair texture, psychosocial traits, public presentation, and status (Sanjek 1971; Harris 1970;

[22] The IBGE (Instituto Brasileiro de Geografia e Estatística) has traditionally used a racial/colour question in demographic surveys and censuses (Nobles 2000). In the 2010 census, most of the Brazilian population was still divided into three main colour/racial categories (*branco*, *pardo*, and *preto*) which comprise over 98% of citizens. The categories *amarelo* and *indígena* are less statistically significant. This division has remained substantially similar since 1872 (Petrucelli 2007).

Harris and Kottak 1963; Kottak 1983). This supposedly gives rise to a range of different terms such as *café com leite* (coffee with milk); *cor de mel* (colour of honey); *cor de formiga* (colour of an ant); *cabo verde*; *sarará*; *marrom bombom* (brown sweet); and *bronzeado* (sun-tanned). It has been argued that the preference is typically for descriptive terms that associate a person as much as possible with whiteness and/or race mixture, rather than with blackness. This is because the meaning of blackness is rooted in similar historical experiences, discourses, and practices across the region that devalue Africa, African physical features, and African cultural forms.

Anti-racist organizations in Brazil see a problem in some Brazilians not wanting to identify as black and advocate for merging the Brazilian census categories *pardo* (brown, mixed race) and *preto* (dark-black) into one category, *negro* (black). However, other scholars, such as Sheriff (2003), have argued that it should not be assumed that Brazilians' supposed preference for deploying a range of colour terms when describing people is incompatible with black consciousness or more politicized expressions of blackness. Referring to research she conducted on race and racism in Rio de Janeiro, Sheriff (2003:102) argues that race is conceptualized both as a different and "deeper" quality, and as a simple bipolar category. Perry (2013, xix), drawing from her research in Salvador on black women in urban social movements, argues that Brazilians have no problem identifying with blackness when they "feel" and "see" racism *na pele* (in the skin). In this book, I do not always navigate these terminological issues perfectly. I use the terms Afro-/African descendant, Afro-Brazilian, and black to include both *pardo* (brown, mixed race) and *preto* (black) categories on the national census, in alliance with black activists' struggles to challenge the denial of race in Latin America.

Finally, I would like to call attention to the term "middle-class," the understanding of which could be very different in Brazil as compared to Europe or North America—and even between different regions in Brazil. I explain how I use the term in the context of my research in greater detail in Chap. 1—for now, it is important to note that while the term *classe média* (middle class) was used in everyday language, lower-class residents of Salvador often grouped them along with the wealthy or rich. When quizzed further about this topic in interviews, people from a higher socio-economic status often made finer distinctions, such as *classe média alta* (high middle class); *classe média para média alta* (middle class going to high middle class); *classe média média* or *classe média mesmo* (very middle class); and *classe média baixa* (lower middle class). These ideas surrounding middle-class status often drew on fine distinctions such as ability to speak correct Portuguese, neighbourhood residence, type of house/apartment that one lived in, dress and self-presentation, type of education, and exposure to travel, among others.

A Note on Language

In addition to preparing for fieldwork by taking language classes, I worked with an experienced language teacher in the first few months of research in Salvador. She turned out to be an invaluable research assistant. She assisted with the formulation of questions, going through recorded interviews, finding contacts, and double-checking information. All interviews were conducted in Portuguese and then transcribed and translated either by the author or with paid research assistance done in a confidential manner. I often asked interviewees to re-read their transcribed interviews in Portuguese, to check that I captured what they wanted to say. I take full responsibility for the work that is reflected here.

References

Andrews, George Reid. 2004. *Afro-Latin America, 1800–2000.* Oxford/New York: Oxford University Press.

Anzaldúa, Gloria E. 1981. La Prieta. In *This Bridge Called My Back: Writings by Radical Women of Colour*, ed. Cherríe L. Moraga and Gloria E. Anzaldúa. New York: Kitchen Table, Women of Colour Press.

———. 1990. *Making Face, Making Soul/Haciendo Caras: Creative and Critical Perspectives by Women of Colour.* San Francisco: Aunt Lute Foundation Books.

Azevedo, Thales de. 1996 [1953]. *As Elites de Cor numa Cidade Brasileira.* Salvador: Editoria da Universidade Federal da Bahia.

Basch, Linda, Nina Glick-Schiller, and Christina Szanton-Blanc. 1994. *Nations Unbound: Transnational Projects, Post-Colonial Predicaments and Deterritorialized Nation States.* Langhorne: Gordon and Breach.

BBC News. 2021. Why Colombia's Protests are Unlikely to Fizzle Out. 31 May. https://www.bbc.com/news/world-latin-america-56986821. Accessed 16 June 2021.

Borda, Sandra. 2021. Why Colombia Has Erupted in Protest. *Open Democracy*, 2 June. https://www.opendemocracy.net/en/democraciaabierta/colombia-protest-why-erupted-en/. Accessed 16 June 2021.

Brodkin, Karen. 1999. *How Jews Became White Folks and what that says about Race in America.* Piscataway: Rutgers University Press.

Caldwell, Kia Lilly. 2007. *Negras in Brazil: Re-envisioning Black Women, Citizenship, and the Politics of Identity.* New Jersey/London: Rutgers University Press.

Ciconello, Alexandre. 2008. *The Challenge of Eliminating Racism in Brazil: The New Institutional Framework for Fighting Racial Inequality.* Oxford: Oxfam International. https://oxfamilibrary.openrepository.com/bitstream/handle/10546/112382/fp2p-cs-challenge-elminating-racism-brazil-140608-en.pdf;jsessionid=2F506AA25D7FE52C4BF71A377D03CE60?sequence=1. Accessed 25 July 2021.

DIEESE. 2013. *Os Negros No Trabalho.* São Paulo: Departamento Intersindical de Estatística e Estudos Socioeconômicos. https://www.dieese.org.br/analisepe d/2013/2013pednegrosmetEspecial.pdf. Accessed 16 June 2021.

Diversi, Marcelo, and Claudio Moreira. 2009. *Betweener Talk: Decolonizing Knowledge Production, Pedagogy, and Praxis.* New York: Routledge.

Ferreira, F.H.H., Julian Messina, Jamele Rigolini, Luis Felipe Lopez-Calva, Mana Ana Lugo, and Renos Vakis. 2013. *Economic Mobility and the Rise of the Latin American Middle Classes.* Washington, DC: The World Bank. https://openknowledge.worldbank.org/bitstream/handle/10986/11858/9780821396346.pdf?sequence=5&isAllowed=y. Accessed 16 June 2021.

Figueiredo, Angela. 2003. A Classe Média Negra não vai ao Paraíso: Trajetórias, Perfis e Negritude entre os Empresários Negros. PhD diss., Universidade do Rio de Janeiro, Rio de Janeiro.

Frankenberg, Ruth. 1997. *Displacing Whiteness: Essays in Social Change and Cultural Criticism.* Durham: Duke University Press.

Frazier, Franklin. 1942. The Negro Family in Bahia, Brazil. *American Sociological Review* 7 (4): 464–478.

Freire, German, Carolina Diaz-Bonilla, Steven Schwartz Orellana, Lopez, Jorge Solar López, and Flavia Carbonari. 2018. *Afro-Descendants in Latin America: Toward a Framework for Inclusion.* International Bank for Reconstruction and Development/World Bank. http://documents1.worldbank.org/curated/en/896461533724334115/pdf/129298-7-8-2018-17-29-37-AfrodescendantsinLatinAmerica.pdf. Accessed 16 June 2021.

Guimarães, A.S.A. 1987. Estructura e Formação das Classes Sociais na Bahia. *Novos Estudos, CEBRAP* 18 (2): 57–69.

———. 2013. Black Identities in Brazil: Ideologies and Rhetoric. Working Paper Series 52. Berlin: desiguALdades.net International Research Network on Interdependent Inequalities in Latin America. Accessed 25 July 2021 at: https://www.academia.edu/11355363/Black_Identities_in_Brazil.

Haraway, Donna. 1988. Situated Knowledges: The Science Question in Feminism and the Privilege of Partial Perspective. *Feminist Studies* 14 (3): 575–599.

Harris, Marvin. 1970. Referential Ambiguity in the Calculus of Brazilian Racial Identity. *Southwestern Journal of Anthropology* 26 (1): 1–14.

Harris, Marvin, and Conrad Phillip Kottak. 1963. The Structural Significance of Brazilian Categories. *Sociologica* 25 (3): 203–208.

Herskovits, Melville J. The Negro in Bahia, Brazil: A Problem in Method. *American Sociological Review* 8 (4): 394–404.

hooks, bell. 1981. *Ain't I a Woman: Black Women and Feminism*. Boston: South End Press.

———. 1990. *Yearning: Race, Gender, and Cultural Politics*. Boston: South End Press.

———. 1992. *Black looks: Race and Representation*. Boston: South End Press.

Hordge-Freeman, Elizabeth. 2015. *The Colour of Love: Racial Features, Stigma and Socialization in Black Brazilian Families*. Austin: University of Texas Press.

IBGE. 2013. *Pesquisa Nacional por Amostra de Domicílios: Sintese de Indicadores, 2013*. 2nd ed. Rio de Janeiro: IBGE. https://biblioteca.ibge.gov.br/visualizacao/livros/liv94414.pdf. Accessed 16 June 2021.

Kottak, Conrad Phillip. 1983. *Assault on Paradise: Social Change in a Brazilian Village*. New York: Random House.

Landes, Ruth. 1947. *The City of Women*. Albuquerque: University of New Mexico Press.

Lima, Rafael. 2020. How Brazil's Anti-Racism Agency Became a Part of the Problem. *The Brazilian Report*, July 6. https://brazilian.report/society/2020/07/06/brazil-anti-racism-agency-became-part-of-the-problem/. Accessed 16 June 2021.

Morley, Tonya, and Allison Hagen. 2020. Angela Davis Protests Anti-Racism. In *Here and Now*, June 19. Boston NPR News Station, WBUR. https://www.wbur.org/hereandnow/2020/06/19/angela-davis-protests-anti-racism. Accessed 16 June 2021.

Narayan, K. 1993. How Native Is A "Native" Anthropologist? *American Anthropologist* 95: 671–686.

Ncube, M., and C.L. Lufumpa. 2015. *The Emerging Middle Classes in Africa*. Abingdon/New York: Routledge.

Neri, Marcelo Côrtes. 2008. *The New Middle Class*. Rio de Janeiro: Fundação Getúlio Vargas/Centro de Politicas Sociais. https://www.cps.fgv.br/ibrecps/M3/M3_MidClassBrazil_FGV_eng.pdf. Accessed June 16, 2021.

Nobles, Melissa. 2000. *Shades of Citizenship: Race and the Census in Modern Politics*. Stanford, CA: Stanford University Press.

Organization of American States. 2021. IACHR Requests Authorization to Conduct A Working Visit to Colombia in the Wake of Alleged Human Rights Violations During the Social Protests. Organization of American States Media Center, May 14. http://www.oas.org/en/IACHR/jsForm/?File=/en/iachr/media_center/PReleases/2021/125.asp. Accessed 16 June 2021.

Perez-Sarduy, P., and Jean Stubbs. 1995. *No Longer Invisible: Afro-Latin Americans Today*. London: Minority Rights Group.

Perry, K.K.Y. 2013. *Black Women Against the Land Grab: The Fight for Racial Justice in Brazil.* Minneapolis/London: University of Minnesota Press.

———. 2020. The Resurgent Far Right and the Black Feminist Struggle for Social Democracy in Brazil. *American Anthropologist* 122 (1): 157–162.

Petrucelli, J.L. 2007. *A Cor Denominada: Estudos sobre a Classificação Étnico-Racial.* Rio de Janeiro: DP&A.

Phillips, Tom. 2020. Brazil Official Recorded Calling Black Rights Movement Scum. *The Guardian*, 3 June. https://www.theguardian.com/world/2020/jun/03/brazil-sergio-camargo-black-rights-movement-scum-recording. Accessed 16 June 2021.

Pierson, Donald. 1942. *Negros in Brazil: A Study of Race Contact at Bahia.* Chicago: The University of Chicago Press.

Rahier, J.M., Percy C. Hintzen, and Felipe Smith. 2010. *Global Circuits of Power: Interrogating the African Diaspora.* Urbana: University of Illinois Press.

Sanjek, Roger. 1971. Brazilian Racial Terms: Some Aspects of Meaning and Learning. *American Anthropologist* 73: 1126–1143.

Sansone, Livio. 2003. *Blackness Without Ethnicity: Constructing Race in Brazil.* New York/Houndsmills/Basingstoke/Hampshire: Palgrave Macmillan.

Santana, Ivo de. 2009. À Margem do Centro: Ascensão Social e Processos Identitários entre Negros de Alta Escalão—O Caso de Salvador. PhD diss., Universidade Federal da Bahia: Salvador.

Sheriff, Robin E. 2001. *Dreaming Equality: Colour, Race and Racism in Urban Brazil.* New Brunswick: Rutgers University Press.

Thomas, D.A., and M. Kamari Clarke. 2013. Globalization and Race: Structures of Inequality, New Sovereignties, and Citizenship in a Neoliberal Era. *Annual Review of Anthropology* 42: 305–325.

Twine, France W., and Jonathan Warren, eds. 2000. *Racing Research, Researching Race: Methodological Dilemmas in Critical Race Studies.* New York: New York University Press.

United Nations. International Decade for the People of African Descent 2015–2024. https://www.un.org/en/observances/decade-people-african-descent. Accessed 17 June 2021.

United Nations Digital Library. 2020. A Global call for Concrete Action for the Elimination of Racism, Racial Discrimination, Xenophobia, and Related Intolerance and the Comprehensive Implementation of and Follow-Up to the Durban Declaration and Programme of Action. https://digitallibrary.un.org/record/3896183?ln=en. Accessed 16 June 2021.

Vargas, J.H.C. 2016. Black Disidentification: The 2013 Protests, Rolezinhos, and Racial Antagonism in Post-Lula Brazil. *Critical Sociology* 42 (4–5): 551–565.

Wade, Peter. 2010. *Race and Ethnicity in Latin America.* 2nd ed. New York/London: Pluto Press.

Wade, Peter., James Scorer, and Ignacio Aguiló, eds. 2019. *Cultures of Anti-Racism in Latin America and the Caribbean*. London: University of London.

Wagley, Charles. 1952. *Race and Class in Rural Brazil*. Paris: UNESCO.

Zhang, L. 2010. *In Search of Paradise: Middle Class Living in a Chinese Metropolis*. Ithaca/London: Cornell University Press.

CHAPTER 1

Towards an Ethnography of Upwardly Mobile Afro-Brazilians Living in the City of Salvador

The idea that racial mixture is the essence of Brazilian identity has survived almost intact until recently, even though a number of scholars and black activists have demonstrated the impact of race on Brazilian inequalities since at least the 1950s.[1] The Brazilian ideology of racial democracy celebrates processes of racial and cultural mixing, understood to be rooted in the country's history. This ideology, largely attributed to the Brazilian sociologist Gilberto Freyre (1956, 1961), promotes the idea that national racial divisions had been overcome by the miscegenation of African slaves, Portuguese colonizers, and the indigenous population. Racial democracy is not just an ideology embraced and promoted by national elites but enacted by Brazilians themselves in their daily discourses and practices—a process which is well-documented in the literature on race in Brazil (Da Costa 2014; Sue 2013; Telles 2004; Twine 1998; Wade 2005). Outside of Brazil, the country symbolized the hope of achieving an egalitarian, post-racial citizenship, especially after World War II and the atrocities of Nazi racism in Europe.

The continued importance of this national ideology was made clear to me when I visited the Brazilian Embassy in Kingston, Jamaica, in July 2005 to apply for a visa to carry out doctoral research in the north-eastern city of Salvador. I was informed that I needed to submit certain

[1] Some of the key studies include Fernandes (1969), Hasenbalg (1979), Do Valle Silva (1985), Nogueira (1985), and Costa-Pinto ([1953] 1998).

D. J. Gordon, *Blackness and Social Mobility in Brazil*,
https://doi.org/10.1007/978-3-030-90765-5_1

documents and a short description of my research. I had received a letter of my official attachment to the Centre for Afro-Oriental Studies at the Federal University of Bahia, so I put together a few paragraphs on the research to be undertaken. I explained that my goal was to explore ethnographically the everyday lives, experiences, and social networks of Afro-Brazilians who had become upwardly mobile in a country with persistent racial and economic inequality. Indeed, researchers writing on race relations in Brazil had noted that since the beginning of the 1990s, there were growing numbers of Afro-Brazilians occupying middle-income occupations such as teaching and education, public sector work, the liberal professions, engineering, small business, and white-collar work (Burdick 2005; Figueiredo 2003, 2004; Santana 1999, 2009). Burdick (2005, 329), citing data from the national household survey (Pesquisa Nacional por Amostra de Domicílios, or PNAD), underscored that the number of Afro-Brazilian households earning more than five times the minimum salary nearly doubled between 1992 and 2001. A shift in the educational achievements of Afro-Brazilians in recent decades has also been noted, suggesting the possibility of their entrance into the job market at a more qualified and higher level (Hasenbalg and Do Valle Silva 1991; Telles 2004; Figueiredo 2003). However, new quantitative studies had identified an "elitist profile" for racial inequality in Brazil, the sense that it is more strongly felt by those in higher positions, especially as they tend to live and work in "whiter" social spaces (Campante et al. 2004; Lovell 1999; Mitchell-Walthour 2017). Focusing on socially ascendant Afro-Brazilians would therefore provide another angle from which to examine how they cope with, negotiate, or resist structures of race and class in their society.[2]

I submitted the required documents to the Brazilian Embassy and on September 15th, 2005, I was asked to attend a meeting with the Consul. On the day of the meeting, the secretary ushered me into his office, which had a large glass window offering a panoramic view of Jamaica's commercial capital, New Kingston, with the famous blue-green mountains in the background. However, the expression on his face was not as pleasant.

[2] Versions of this argument are developed in a previous publication by the author, "Negotiating Inequality: The Contemporary Black Middle Classes in Salvador, Brazil," in *People, Money and Power in the Economic Crisis: Perspectives from the Global South*, edited by Keith Hart and John Sharp (2015, New York and Oxford: Berghahn Books, Chap 5, pp. 106–128).

Presiding behind a broad, mahogany desk was a stern looking man. He let out a heavy sigh, "Are you studying at an American university?" he asked. I explained that I was studying at the University of Manchester in the United Kingdom and that I would be affiliated with the Federal University of Bahia. Looking like he was about to burst a blood vessel, he declared that, "We do not have black people in Brazil. Brazilians are a mixed-race people." With a dismissive wave of his hand in the direction of the blue-green mountains behind him, he continued, "Here in Jamaica, you have blacks, your country is full of them. But in Brazil people mix. Everybody is mixed, we are a mixture of European, Indian and African." Keeping quiet proved to be an effective strategy throughout the Consul's telling of the history of Brazil, which was of a country uplifting itself through race mixing. Bahia seemed to be an especially problematic location in the Brazilian landscape for him, upsetting cherished notions of nationhood grounded in ideas of racial blending. He warned that Bahians (especially the black activists there) tended to emphasize and exaggerate the importance of the black contribution to Brazilian people and culture. Shaking his head, he stated that, "It is more like what Gilberto Freyre, our great sociologist, said about Brazil. He was the first sociologist to eschew racism and give all the races their proper place." He argued that different parts of Brazil were composed of different kinds of people. For example, he referred to the more European nature of the South; that the North had more indigenous people; and that the Northeast had more black people. He exclaimed that the people from Bahia loved to dance and sing but could not be trusted in terms of the quality of their work. With that pronouncement, he directed me to the universities in São Paulo and Rio de Janeiro which he believed were more objective and rigorous. Suddenly, he smiled broadly and said that although he was not in agreement with my topic, he wasn't a dictator and that he would issue the visa.[3]

The Consul's denial of the existence of Afro-Brazilian people came as culture shock to me, as in my Jamaican social milieu, there is strong debate around issues of race, identity, and slavery. Of course, this debate is not absent in Brazil, where anti-racist mobilization as well as the implementation of redistributive policies have shaped the public discourse on race and

[3] Due to ongoing research in Brazil, I continue to interact with the Brazilian Embassy in Jamaica. My experience has been that staff are generally helpful and friendly. They have also greatly assisted me with translation services.

racial inequality in recent years.[4] Indeed, the framing of blackness and anti-racist politics as foreign to Brazil (both in popular and academic discourse inside and outside of the country) not only reflects a lack of understanding of Brazil's black radical tradition, it also demonstrates how a narrow, local lens can obfuscate similarities and differences in the experiences of African diaspora communities globally.

The black middle classes in Brazil, though small in number, have managed to carve out strategic positions within the state and some have adopted a more political racial identification as *negro* (black), that embraces the national census categories of *pardo* and *preto* and that stands in opposition to white (Silva and Reis 2011). Thus, contemporary black identity politics in Brazil works to counteract the ways in which ideologies of race mixture have worked to minimize racism. However, my experience with the Brazilian Consul caused me to think more critically about Brazilian race dynamics. For example, I came to appreciate that there are real aspects of peoples' lives that underwrite their belief in racial democracy—such as the existence of inter-racial marriages, moments of shared conviviality, and friendships between different ethnic groups. Furthermore, a traditional preference for *racismo cordial* (cordial racism) among Brazilians, by which individuals tend to downplay racial differences in social interactions that might lead to conflict or disagreement, might explain why some of my research participants, regardless of skin colour, shied away from anti-racist or pro-black rallying calls in street protests or parades.[5] Yet structural racism has always existed in Brazil, embedded in social, economic, and political institutions in such a way that privileges associated with "whiteness" and disadvantages associated with "blackness" have endured over time.

Anthropologist Jennifer Roth-Gordon highlighted the fact that racism can co-exist with racial mixture and conviviality in Brazil, a situation which she refers to as Brazil's "comfortable racial contradiction" (2017, 5). Yet, as Roth-Gordon pointed out, this racial situation is not "comfortable" for all Brazilians, nor is it an example of how Brazil developed a set of

[4] Twenty-first century discourses on race in Brazil and Latin America more generally have had varying effects on the recognition of racism in different countries and sparked debates about policies to address racial injustice. See Bailey (2009), Golash-Boza (2010), Guimarães (2006), and Silva and Reis (2012).

[5] For a discussion on cordial racism, see for example Fry (1995–1996) and Turra and Venturi (1995). A similar cultural practice, the *jeitinho* (or "little way," which includes granting favours or bypassing rules or laws), illustrates the Brazilian preference for avoiding conflict (Barbosa 1992).

"exceptional" racial beliefs. Indeed, people from across the African diaspora juggle opposing ideals, discourses, and realities as they try to make sense of the paradoxes of unequal society. For example, in Jamaica our national motto is "Out of Many, One People," yet ordinary black Jamaicans and poor communities disproportionately bear the brunt of the government's fight to control rising crime and violence. They are no strangers to the policing of their neighbourhoods and to the extra-judicial killing of black people—just as in countries such as Trinidad and Tobago, Brazil, the United States, and South Africa. It is imperative that we understand the wider structural dynamics of race across the black diaspora, the mutually constituted categories of difference that shape our social worlds and their cultural and economic impacts.

During my first few encounters with Brazilians in Salvador, my proposed topic elicited varied responses. A taxi man with a weathered face exclaimed that there are only *os ricos* (the rich) and *os pobres* (the poor). For him, there were certainly no black middle classes, in a city largely neglected in Brazil. In another conversation with a professor at the Federal University of Bahia, she argued that in Salvador, "*a classe media negra não existe!*" (a black middle class does not exist!). She stated that only white people were middle class because only they had the material base to continue reproducing their status. Meanwhile, black people who became upwardly mobile usually carry the burden of their family members' welfare on their back, thus hindering them from establishing themselves firmly within the middle classes. Yet another Brazilian with whom I spoke, a black activist and director of a non-governmental organization, firmly identified as black middle class and stated that the work of his organization was to focus on young black Brazilians and help them to get into public universities which have long been associated with white privilege. In doing so, he hoped to consciously create an educated, professional group of black middle-class people in the city.

These first encounters during fieldwork impressed on me the politicized nature of race in Brazil and the need to sometimes take a more nuanced approach, allowing "race" to emerge in conversations and interactions with Brazilians. I kept a dairy that detailed my personal experiences and observations of daily interactions and paid attention not only to what people said, but also to what people did in everyday social situations. I used these personal reflections in conjunction with my formal notes and interview materials throughout the research process. In addition, I adopted a flexible approach to ethnographic fieldwork, living and socializing in different neighbourhoods and talking to people of different age, gender, and socio-economic backgrounds. This allowed me to cross lines of class and skin colour, revealing a more complex, heterogenous picture of social stratification and mobility in the city.

Race, Racism, and Anti-racism

Social scientists have historically argued that race would eventually give way to class as the defining factor of social position in a rapidly industrializing Brazil. However, issues of race and racism have hardly faded away. Black activists have argued that Brazilian racial democracy effectively smooths over deeply troubling issues, such as the selective celebration of certain Afro-Brazilian traditions by the state while at the same time, black people are being terrorized and murdered by the state as casualties in the fight against crime and violence (Vargas 2016; Smith 2016).[6] The widely publicized assassination by paramilitary forces of Marielle Franco—an elected politician and human rights activist who strongly opposed the treatment of poor, black people in Brazil—in Rio de Janeiro on March 14th, 2018, followed by the death of João Alberto Silveira Freitas, a welder beaten by white security guards at a supermarket chain in Porto Alegre on November 20, 2020 (on the Day of Black Consciousness in Brazil), have brought issues of race, racism, class, gender, and sexuality back into the national and global spotlight once more. Given this current context and the significance of these issues for Afro-Brazilians and for the black diaspora more generally, there is an increased focus on naming racism, uncovering it, and calling it out. Therefore, it is worth clarifying the terms race, racism, and anti-racism here. Nevertheless, there is a risk that focusing on these labels may cause us to ignore the variety of ways in which social and racial inequalities, both material and symbolic, are reproduced, challenged, and disrupted.

[6] An analysis of homicides and sexual violence in Bahia showed that in 2018, 5427 black men were killed, while the number of white men killed was 350. Furthermore, black women were two times more likely to be sexually assaulted than white women in Bahia. See the report, 'A Cor de Violência na Bahia – Uma Análise dos Homicídios e Violência Sexual na Ultima Década,' by Rede de Observatórios da Segurança, 2020, pp. 4–6. http://observatorioseguranca.com.br/produtos/relatorios/. Accessed July 23, 2021.

Today, most scholars reject the concept of biological race and agree that it is a social, historical, and political formation.[7] Viewing "race" as a social construction, however, does not mean that "race" is not real. This approach resonates with the view that, "race, like gender, is 'real' in the sense that it has real, though changing, effects in the world and [a] real, tangible, and complex impact on individuals' sense of self, experiences, and life chances"(Frankenburg 1993, 11). It is important to note that social constructivist approaches to race and other forms of identities have been increasingly critiqued, with some scholars proposing alternative views that recognize the continued importance of both identity and identity politics (Alcoff 2006; Alcoff et al. 2006; Mohanty 1993, 2000; Stephen 2001). Mohanty's formulation of a realist theory of identity points to the need to view identities as both "real" and constructed and draws attention to the epistemic and political significance of identities. Employing a realist theory of identity opens the possibility of exploring how my research participants engaged with, transformed, resisted, and deployed identity categories, without necessarily dismissing their appeals to identity as essentialist.

By the twentieth century, the concept of race had largely been replaced in mainstream Euro-American intellectual circles by that of ethnicity, which focused more on people's ideas about cultural difference, thereby avoiding the negative connotations of the earlier scientific approaches to race (Beckett 2017). This influenced Brazilian social sciences, such that the concept of race came to express the ignorance of those who used the term and racism came to be viewed by some as a foreign invention. Commenting on this issue, the Brazilian sociologist Antonio Sérgio Guimarães stated that it was interesting to see how this anti-racialist ideal was absorbed into the Brazilian way of being, such that it became cliché that Brazilian races do not exist and that what was important in terms of life opportunities was social class (Guimarães 2001). Only in recent years has this idea undergone systematic critique from the Brazilian Black Movement and some social scientists.

A useful way to think about how race and phenotype affect people's daily lives in Brazil is to view race as a process and a product of racialization processes, rather than a status (Hordge-Freeman 2015, 4). Racialization refers to the process by which meanings are assigned to

[7] For a detailed discussion of the history of ideas about race and transformations over time, see Banton (2015), Wade (2010), and Wade et al. (2019).

physical and cultural traits, and it is "produced and reproduced through ideological, institutional, interactive, and linguistic practices that support a particular construction of Difference" (Dominguez 1994, 333). Racialization is important because the meanings that are created and assigned to different bodies sustain racialized systems like that in Brazil (and elsewhere). These developments should invite us to critically consider race, racialization, and racism; to evaluate existing approaches; and to think about how to dismantle racial injustice and its harmful social effects.

A typical dictionary definition of racism is that it refers to a set of negative attitudes that people hold about another perceived ethnic group. Different authors have highlighted varied aspects of the term (Banton 2015; Goldberg 1993; Kottak 2015; Miles and Brown 2003; Wade et al. 2019). For Kottak (2015), racism may be understood as a range of attitudes, beliefs, and behaviours—including active exclusion of blacks from resources and institutions, that is, racial discrimination. Racial discrimination is not the same as other kinds of discrimination, as it is grounded in ideas that link perceived human differences to notions of blood, biology, phenotype, and inherited essences (Wade 2010; Wade et al. 2019). Furthermore, racial discrimination typically involves specific language, sets of stereotypes, and ideas about where people sit in relation to each other in the value and power hierarchies of the world (Wade et al. 2019, 11).

While racial discrimination tends to be associated with individual acts or attitudes, racism trains our gaze towards culturally embedded systems of beliefs and values and pervasive social structures of discrimination and privilege. Racism is therefore reproduced by structural conditions and is harder to change, affecting access to jobs, housing, schooling, and other social conditions (Kottak 2015; Telles 2004). Furthermore, I follow Wade et al. (2019, 2) in seeing racism as a product of histories of colonialism, slavery, and post-colonial power dynamics, as well as inequality within the world order. This point of view is not new, as it builds on strands of writing in colonial and post-colonial studies, including the work of scholars and thinkers such as Eric Williams ([1944] 1994), Walter Rodney ([1973] 1983), and Marcus Garvey (1967).

More recently, coloniality theorists see racism as underpinning concepts and projects such as "modernity" and "civilization" (Mignolio 2011; Quijano 2007; Restrepo and Rojas 2010). These approaches share commonalities with today's critical race perspectives—especially the insight that racism is an integral part of the global order, structured by a colonial past and post-colonial relations of domination (Thomas and Clark 2013;

Goldberg 2008). Thus, racism is understood to be "a system that distributes power, privilege, welfare and security among people distinguished, classified, and stratified in terms of ideas about their physical nature and their behaviour" (Wade et al. 2019, 2).

However, the use of the terms racism and racial discrimination has been uneven across Latin America; indeed, other terms may have been used (see Wade et al. 2019, 8–9). The Brazilian social scientist, Oracy Nogueira (1985), has argued that Brazil was characterized by *preconceito de marca* (racial prejudice based on physical appearance) as opposed to *preconceito de origem* (racial prejudice based on descent). Indeed, there is the idea that racism does not exist in Brazil, only prejudice, meaning mistaken individual perceptions, which tend to be corrected in the dynamic context of ongoing social relations (Guimarães 2001, 158).

In recent times, so-called cultural racism has been identified in the wider scholarly literature on race, which relies on culturally based arguments that are often linked to older conceptions of race (Ryan 1976; Stolcke 1995). Thus, perceived racial types (such as "whites," "blacks," "Indians," and "Asians") continue to be identified and to be discriminated against, but now based on their "culture," rather than their biology. Although explicit references to biology or race are played down in this type of discourse, there is a sense in which culture is naturalized and seen as part of a person's or group's nature, passed down from one generation to the next (Bonilla-Silva 2018; Wade 2010; Wade et al. 2019). For example, in Latin American and Caribbean contexts, whiteness and non-whiteness are often discussed as differences in the amount of culture one has, which implies that social and racial differences are closely intertwined, though they do not always map unto each other. In Brazil, this idea is captured in phrases such as *gente sem cultura* (people without proper socialization), *mal educado* (badly educated or displaying poor manners), and *brega* (tacky people), in which racialized distinctions are made between those "with" culture and those "without" culture. These discourses play on the symbolic association of whiteness to civility and discipline and blackness to lack thereof, while at the same time not necessarily implicating phenotype. In other words, being well-educated, well-spoken, possessing wealth, being a professional, or being well-travelled is linked to whiteness, although such a person could be light or dark-skinned. Thus, cultural racism is one way in which racial hierarchies can be justified without a direct reference to race. Latin American and Caribbean contexts have had long experience with this type of racialized discourse, which has its equivalent in so-called post-racial societies as "colour-blind racism" (see Bonilla-Silva 2018).

Regardless of the angle or lens through which concepts of race and racism are examined in the world today, it is impossible to escape its reach. Both racism and anti-racism have become more widespread and explicit, following earlier developments (e.g. the turn to multiculturalism in the 1980s) and more recent occurrences, such as the spread of popular nationalism and white supremacy ideologies, the election of Donald Trump, Brexit, and the emergence of the global Black Lives Matter movement. In Latin America, racist stereotypes and caricatures are increasingly being challenged (Wade et al. 2019, 1–2). Furthermore, indigenous people and African descendants across the Americas are mobilizing to form social movements and organizations, demanding that their humanity be recognized.

Transformations such as these have helped to shift the discourse on race in Brazil, such that racial inequality and racism are increasingly identified and recognized by the Brazilian population. However, a recent opinion poll carried out by Perseu Abramo Foundation in 2003 stated that while 87 per cent of all Brazilians admit that there is racism in Brazil, only 4 per cent of them acknowledge themselves to be racists.[8] This can lead us to two conclusions: firstly, that racism exists not because of the awareness of the perpetrator, but because of its effects on those who suffer from it; and secondly, that racism in Brazil is always blamed on other people and never on the daily practices of its agents, making it difficult to eradicate. Nevertheless, recent developments (such as Black Lives Matter protests across several Brazilian cities) suggest that citizens are digging deeper in their understandings of white privilege and structural racism—even within the present context of a wave of Far Right politics shaping the country.[9] Having outlined the concepts of race, racism, and anti-racism that are key to this book, I will now return to my ethnography to show how these issues play out in contemporary urban Brazil.

[8] These figures were cited in Ciconello (2008, 1), who quotes an opinion poll carried out by Perseu Abramo Foundation in 2003.

[9] See Daniela Gomes, 'Brazilian Politics and the Rise of the Far Right.' *Black Perspectives*, 7 January 2019. https://www.aaihs.org/brazilian-politics-and-the-rise-of-the-far-right/. Accessed July 24, 2021.

Studying the Urban Black Middle Classes

My first meeting with Jorge Carvalho took place at his business office in Iguatemi, a commercial neighbourhood regarded as very modern and cosmopolitan, as reflected in the architecture of the area, the organized streets and services, the impressive skyscrapers, and newer commercial buildings. Jorge described himself as a self-made businessman, originally from a small town about 100 km from Salvador in the Recôncavo area of Bahia. In his 50s, he had migrated to the capital city as a young man, carving out a successful career for himself in a major construction firm. Jorge lived in the quiet and wealthy neighbourhood of Graça. He eventually put me in touch with a friend of his, Túlio Santana, who worked as the Executive Director of a transport syndicate in a glass-and-steel skyscraper located in Iguatemi. A well-travelled engineer, Túlio lived with his family along the Avenida Atlântica in a modern, secure apartment with a beautiful view of the coast.

Jorge and Túlio's location in a powerful commercial district, occupying senior level management positions and living in *bairros nobres* ("noble" neighbourhoods), may have aligned them with whiteness in the eyes of others—that is, they had acquired the symbolic trappings of the privileged Euro-Bahian elite through upward social mobility in Salvador's social hierarchy. Contrary to popular and academic discourse that such upwardly mobile Afro-Brazilians would likely have an ambiguous relationship with racial identity, Jorge and Túlio strongly identified as black and were both involved in various Black Movement and Afro-Brazilian cultural activities across the city. They recognized that racial problems in Brazil are the result of several generations of systemic social and economic exclusion, negatively impacting black people throughout Brazil. However, they tended to describe systemic racism as something that mainly affected the lives of poorer Afro-Brazilians rather than as significantly impacting on their own lives.

One Saturday morning, Jorge invited me for lunch at his luxurious penthouse apartment which he shared with his wife. From the balcony of the apartment, he pointed out the families who inhabited some of the private mansions in the area as well as other buildings in the neighbourhood. I surveyed what I could not see from the ground level—the swimming pools, tennis courts, mini golf-courses, and private gardens located on the roof tops of older and newer apartment buildings, as well as the menacing security guards protecting the residents within, standing silently with machine guns tucked closely against their sides. I wondered if Jorge and Túlio were ever invited over to dinner by their wealthier neighbours. I glanced at a framed picture of Jorge posing with Jesse Jackson, the African American politician, on a visit to Bahia. Had Jorge and Túlio sought out connections in the Afro-Brazilian and black diaspora community by personal choice or as a result of exclusionary practices in the primarily "white" social circles in which they circulated? What would they think of the comments of another highly positioned Afro-Brazilian whom I interviewed, a senior level employee of the Central Bank in Salvador, when he stated that even though he had done everything "according to the book," he felt that his presence was only tolerated in some places?

While Jorge and Túlio described incidences of racial discrimination that they faced in their daily lives, they were not of the view that these experiences hampered their overall climb to success. Although they both subscribed to some ideas about racial democracy, Jorge and Túlio appear to be motivated by a desire to exert some control over the terms upon which black participation in their society is organized. Jorge expressed the view that Afro-Brazilians were on the outside of the realms of economic and financial power and therefore needed to speak more openly about the problem of financial powerlessness or else everything would continue as before. He stated:

> Power has its proper rules, and we don't know what they are. They are constantly in motion, always changing, to maintain the process of power and the people who have it. When we (blacks) finally find out the answers, the questions have already changed. So, we are always out of power's sphere. (Interview: November 24, 2006)

Their perception is like what researchers have reported. For example, the Brazilian Ethos Institute has carried out several studies since 2001 on the social, racial, and gender profile of the 500 largest companies in Brazil. In their 2016 report, it was noted that a hierarchical bottleneck existed for Afro-Brazilian personnel, as was the case in previous studies, with Afro-Brazilians becoming increasingly underrepresented at higher levels.[10] Among the 117 companies surveyed in 2015, Afro-Brazilians represented

[10] Note that surveys were conducted in 2001, 2003, 2005, 2007, 2010, and 2015. Claudio Pucci, *Social, Racial and Gender Profile of the 500 Largest Brazilian Companies* (São Paulo: Instituto Ethos, 2016). https://publications.iadb.org/en/social-racial-and-gender-profile-500-largest-brazilian-companies. Accessed September 27, 2020.

only 4.9 per cent of directors, 4.7 per cent of executives, 6.3 per cent of managers, and 25.9 per cent of supervisors.[11] The situation was far worse for Afro-Brazilian women, with only 2 Afro-Brazilian women represented among the 548 executive directors included in the study.[12] Furthermore, the vast majority of companies said that they had not taken measures to encourage or enhance the presence of Afro-Brazilians at any staff level, despite the fact that many chief executives perceive that the share of Afro-Brazilians at the manager and executive levels is well below what it should be.[13] Thus the racial framework in Brazil maintains privileges and feeds social exclusion and inequalities.

Recognizing that society is skewed against blacks, Jorge became involved in mentoring younger Afro-Brazilians through the Association of Afro-Brazilian Businessmen, an organization he helped to establish in order to support recent university graduates who needed guidance in their careers. In some cases, he helped them to set up small businesses, particularly those focusing on niche products for the tourist market or products aimed at a black consumer market. Jorge felt that black people needed to look out for each other in a challenging environment. Nevertheless, he was hopeful and optimistic about the future of Afro-Brazilians, especially with the recent implementation of racial quotas in public universities.

The examples of Jorge and Túlio demonstrate that there is a phenomenon of being black and middle class in contemporary Salvador that needs to be analysed, despite denial that it exists. The presence of highly positioned African descendants in the social hierarchy of Salvador has often been a key plank of claims to racial democracy. However, the contemporary black middle classes appear to be emerging in ways that challenge popular, long-standing ideas in the literature and in public discourse about the "whitening" and assimilation of upwardly mobile black Brazilians. The example I outline above reveals an alternative and contradictory narrative, hinting at greater complexity and ambiguity underlying processes of social mobility and the relationship to racial identification.

[11] Ibid., 22. In comparison, white Brazilians comprised 95.1 per cent of director positions; 94.2 per cent of executive positions; 90.1 per cent of managerial positions; and 72.2 per cent of supervisory positions.

[12] Ibid., 23. Afro-Brazilian women represented only 10.3 per cent of workers, 8.2 per cent of supervisors and 1.6 per cent of managers. At the executive level, their share goes down to 0.4 per cent .

[13] Ibid., 23.

In recent years, there is growing interest in the contemporary black middle classes in Brazil. Market researchers are interested in mapping out the size, income levels, and consumption patterns of the "new" black professional classes said to be participating in a more general expansion of the middle classes in Latin America.[14] The tendency is to identify them as consumers sustaining markets for niche products such as lifestyle magazines, cosmetic, and hair products.[15] Others doubt that the black middle class could match the strength and wealth of the lighter-skinned middle class. The question frequently posed is whether the black middle class can challenge the racial hierarchy that has traditionally limited Afro-Brazilians' access to resources and opportunities in Brazil. Black middle classes have lately succeeded in carving out a strategic position within the state. Some see them as crucial actors in the fight to dismantle racial barriers, while others disparage their motives as being merely self-serving. Indeed, it may very well be the case that educated Afro-Brazilians have effectively articulated a political agenda at the expense of disadvantaged groups, including poor black people. However, all such claims require an in-depth understanding of people's lives.

Traditionally, upward mobility for a dark-skinned Brazilian meant aligning oneself with the Euro-Brazilian elite in some way—through kinship, patronage, godparent ties, or adoption. For example, Azevedo's research on *elites de cor* (coloured elites) in Salvador during the 1950s emphasized the evolution of distinct strategies of social mobility resting on the possibility of some degree of integration into the privileged classes through practices of "whitening." These practices might include acquiring money or education, which is symbolically associated with whiteness, or marrying into Euro-Brazilian families (Azevedo [1953] 1996). His account has some basis in reality, without entirely convincing this researcher

[14] Brazil's middle classes are said to be undergoing rapid expansion due to monetary stabilization, economic and industrial growth, the generation of more formal jobs, and availability of flexible credit (see John Parker, "Burgeoning Bourgeoisie," *The Economist*, 14 February 2009, 3–4). However, since about mid-2014, the country has been experiencing an economic recession, the effect of which is still not clear on the so-called rise of the "new" middle classes.

[15] A study conducted at the end of 1997 by the advertisement agency Grottera Comunicação indicated that Brazil's black middle class was seven million strong, with an average monthly family income of BRL $2000 (or roughly US$1168.91). The purpose was to create a profile of the black consumer. For example, every month, this total population is said to spend around BRL $500 million (close to US$3 million) on non-essential products.

that the race and class system was equal and democratic. Guerrero Ramos (1995) was very critical of this group who, he thought, shared the pathologies of whites, especially by accepting their aesthetic values. Carl Degler (1971) proposed the famous “mulatto escape-hatch” theory, when he argued that lighter-skinned Brazilians could easily assimilate into white society, giving up their black identities. He noted that the expectation of interracial mobility had the disadvantage of preventing Afro-Brazilians from attaining group solidarity that would have enabled them to mobilize and fight for equal rights in the way that African Americans did in the 1940s and 1960s. Subsequently, scholars have questioned the notion that modern Brazil has a “mulatto escape hatch” (Andrews 1991; Paixão 2004). Luiz de Aguiar Pinto (1953) differentiated between the old and the new black elites, arguing that, while the old elite would try to pass as white, the new elite were more militant and asserted their black identity, even in the face of racial discrimination. Both the new and the old black elites, however, were said to be disconnected from the large black working class and more likely to organize around class rather than race-based identities, distancing themselves as much as possible from poor, black communities.[16]

My ethnography reveals another aspect to this story in that some upwardly mobile individuals remain selectively involved in their community of origin. In contrast to some scholars, (Figueiredo 2003; De Santana 2009; Da Silva Soares 2004), who emphasized the individual nature of black social mobility, my ethnography reveals that kin support and social networks were important to social ascension. Moreover, actions stressing one’s connections to the Afro-Brazilian “community” (such as sponsoring a school in a poor neighbourhood) have always been a powerful way of achieving status and prestige in Salvador.

Recently, studies of race relations have shifted their focus towards investigating persistent racial inequality, utilizing national statistics to demonstrate that whites fare better than non-whites across a wide range of socio-economic indicators such as income, occupation, access to health and education (Hasenbalg 1985; Castro and Guimarães, 1993; Castro and Barreto 1998; Hasenbalg and Do Valle Silva 1999). Some more

[16] Please see previously published work by the author, “Negotiating Inequality: The Contemporary Black Middle Classes in Salvador, Brazil,” in *People, Money and Power in the Economic Crisis: Perspectives from the Global South*. Edited by Keith Hart and John Sharp (2014, New York and Oxford: Berghahn Books, pp. 106–128).

qualitative approaches have analysed upwardly mobile Afro-Brazilians in relation to their work as activists in non-governmental organizations (Dos Santos 2008; Giacomini 2006); their lifestyle and consumption patterns (Da Silva Soares 2004); the strategies and difficulties of mobility (Figueiredo 2003; De Santana 2009); and their experiences of racial discrimination (Figueiredo 2002; Lima 2001; Silva and Reis 2011).

Angela Figueiredo's writings on the urban black middle classes have probably received the most recognition in Brazil. Focusing on small business owners in Salvador, she argues that "the majority of the [Brazilian] black middle class is part of a first generation that does not have much consolidated economic, social and symbolic capital as most whites in the same class position" (2003, 92). She emphasized the precarity and instability of their position, suggesting their insecurity in relation to the future stabilization of their position in the middle classes. While not discounting Figueiredo's findings, I interviewed research participants who considered themselves (and were seen by others) to belong to high society for generations.

My approach differed somewhat from Angela Figueiredo's. Firstly, my research participants worked in a range of economic sectors, which allowed for a more diverse picture of what the black middle classes are doing in Salvador than if I had concentrated on a single occupational group, such as business-people or professionals. Secondly, I explored the social networks in which black middle-class people are embedded—whereas the methodological focus before tended to be on upwardly mobile individuals. This revealed the heterogeneity of black middle class lives as well as perspectives and attitudes to issues such as community involvement and consumer expenditure.[17]

Studies of the black middle classes in the United States, though not directly comparable to black middle classes in Brazil, also offer some relevant insights. Black middle classes in the United States have been described as economically better off than poorer blacks yet marginalized when compared with the white middle class. Furthermore, they are vulnerable to discriminatory practices in both public and personal life, likely contributing to the formation of exclusive social clubs, activities, and institutions

[17] These arguments were developed in a previous publication by the author, "Negotiating Inequality: The Contemporary Black Middle Classes in Salvador, Brazil," in *People, Money and Power in the Economic Crisis: Perspectives from the Global South*. Edited by Keith Hart and John Sharp (2014, New York and Oxford: Berghahn Books, pp. 106–128).

among the wealthiest African American families (Bowser, 2007; DuBois, 1899, 1903; Frazier, 1957; Graham, 1999; Landry, 1987; Patillo-McCoy 1999). Some of these scholars have also analysed the relationship between middle- and lower middle-class blacks. Issues surrounding black mobility were often described as being highly contentious, frequently involving delicate negotiations, practices of differentiation and solidarity, possible accusations of abandoning poorer relatives, and of taking on the values and lifestyles of the whites. These issues are not entirely different in Brazil. As middle-class Afro-Brazilians negotiate the racial hierarchies of their society and the practicalities of everyday life, they often must be selective about social networks and decide on the familial obligations they are willing to fulfil. These social manoeuvrings open them up to criticism from others and sometimes result in a kind of Janus-faced predicament, where individuals display sharply contrasting characteristics in varied contexts.[18] If ethnographic studies of lower-class and lower middle-class communities in Brazil reveal ambiguity, heterogeneity, and flexibility in racial identification, the expectation is that it probably is not that different as people become upwardly mobile.[19] Writing ethnography in this context meant learning to negotiate potentially tense relationships within families and being sensitive to the contradictions as well as opportunities that social mobility creates for African-descended families.

Conceptual and Theoretical Matters

As a cultural anthropologist, I am trained to see things as a social scientist, using conceptual terms such as "race" and "class" as ways to describe social hierarches, socio-economic structures, and emergent identities. I view race and class not as descriptors of particular groups or persons, but as categories that act as principles of distinction.[20] What these categories mean, how they combine in different ways, and how they affect a person at a given time may vary—they are neither rigid, nor fixed, nor timeless. These concepts appear frequently in this book and are suggestive of larger forces impacting on peoples' lives. However, the interpretive lens that I

[18] I borrow the term "Janus-faced" from discussions of ethnic consciousness and identity in Wade's (2001) work on racial identity and nationalism in Latin America and Comaroff's (1987) work on the nature of ethnicity and its diverse experiential forms.

[19] See, for example, Twine (1998); Sansone (2003).

[20] I derive this from Pierre Bourdieu's (1984) discussion of what he calls "principles of differentiation" in his study of distinction.

use in this book will also direct our gaze to the micro level, on how some of the people I met in Salvador make sense of these forces in their everyday lives.

Theoretical discussions concerning race and class reveal a terrain of dispute. Classic Marxist approaches to race start with the underlying premise of the core division between the owners of capital (the bourgeoisie) and the workers (the proletariat) in capitalist societies, with an emphasis on the exploitation of the workers. This core division determines what happens at other levels of the society so that racial categories are seen to be related to this division. The basic argument is that these divisions exist because the bourgeoisie has created them, either to better dominate a section of the workforce or to create a "divide-and-rule" situation in which the workers are divided into antagonistic racial categories. Thus, racism is embedded in the class relations of colonialism and post-colonialism. Critiques of this kind of argument, however, reject the idea that race is economically determined, pointing to changing racial identifications over time and arguing that these are not easily explained in solely economic terms or in terms of changing class structures (Anthias and Yuval-Davis 1992).

Other scholars have looked at race as an experience and cultural reality, in its own right (Gilroy 1993). As part of the post-modern turn in social sciences and humanities and the subsequent move away from grand theories or totalizing approaches, the emphasis is now on capturing a sense of multiple voices in social life and the different ways in which people may identify difference and sameness, as well as make alliances and enmities. Following this approach, Michael Omi and Howard Winant ([1986] 1994, 2014) developed the racial formation theory, in which they clearly oppose all forms of class reductionism, preferring to analyse what they call "racial formations" and "racial projects." This is one of the more significant, recent attempts to develop an effective and critical race theory that has an adequate historical approach, considering micro- and macro-level social relations as well as racial politics.

At the same time, black people exist within wider economic and political contexts, so they may not find themselves with many options for identification—indeed, they are likely to find themselves coercively identified. Recalling older debates about race, capitalism, and the political economy of Afro-Latin America (e.g. Fontaine 1980), I am concerned with the economic and material processes in which research participants are embedded, thereby challenging newer approaches to social and cultural identities which have tended to submerge or dilute the workings of the economy.[21] However, Stuart Hall, who played a formative role in the development of cultural studies in Britain, has never completely abandoned the force of economics or the material in the shaping of social identities and rejects what he calls the "free floatedness" of more discursive approaches that construe the economy as just another type of discourse (Hall 1997, 25–46). I think it is important to grasp these larger economic questions because changes in material conditions have been shown to influence the ways in which people behave and think. For example, a middle class created by the state, as occurred in many Latin American countries during the 1950s and 1960s, will likely behave and act differently to a middle class based on the private sector and the movement of global finance and job markets.[22] In the wake of the implementation of neoliberal policies in Latin America (and worldwide), markets have been opened up to global trade and state protectionism and welfare have been reduced, pushing down wages for the lower classes and increasing poverty and suffering (Gledhill 2005; Perlman 2004). In this context, Goldberg (2008) insists on recognizing race and racism as vital forces in today's world and sees them as global forces that operate across national boundaries.

The literature on class (mostly from history, sociology, and economics) is equally contentious and difficult to define. For example, do we mean class structure, identity, consciousness, action, and so on when we speak of class? Furthermore, who and what is middle class is a constantly moving target, subject to social, historical, economic, and historical forces (Owensby 1999; O'Dougherty 2002). Nevertheless, class inequality exists beyond its theoretical representation: to think that it does not matter in everyday social life is only a prerogative of those unaffected by the deprivations and exclusions it produces (Skeggs 1997).

Brazilian researchers have noted that quality of life indicators vary from state to state, which means that what it means to be middle class varies across the country. For example, the Northeast has typically lagged the Southeast in terms of average household income (Figueiredo 2003, 46). Most authors agree that middle-class status is based on a combination of

[21] I refer specifically here to Paul Gilroy (1993) and more generally to some branches of cultural studies, postmodernism, and risk society theory that have displayed distinct lack of interest in the economic system of society.

[22] Cohen (2004) illustrates this point for Morocco as she shows how market liberalization has led to the transformation of the role, structure, and consciousness of the middle classes.

factors ranging from socioeconomic indicators such as income, occupation, and education to normative judgements (such as where people live, where they socialize, where they go on vacation, what they wear). Some researchers have also suggested that it is shaped by people's imagination of their position and well-being in the future (Neri 2008; Ortner 2003; Owensby 1999). "Middle class" may be defined in relative terms, as the middle-income range of each country, or in absolute terms, using a fixed band for all countries. Utilizing this definition, Brazil could be said to have *middle classes*: those who are middle class according to global standards (thus constituting a global middle class) and those who are middle class by Brazilian standards. In his study of the "new middle classes" in Brazil, economist Marcelo Neri identified five categories, ranging from A to E, based on calculations of household income. Class C receives the average income in society, with Class D and E representing the poorer classes. Class C is said to be the category that is growing rapidly, with household income of between 1000 reais (US$130) and 4500 reais (US$580) a week. They typically have access to credit and durable goods, such as a television set, washing machine, video player, and a car/motorbike (Neri 2008). The people I interviewed generally had salaries within the range indicated for Class C, although some earned more.

Traditionally, social scientists have tended to rely heavily on income, occupation, and patterns of consumption when defining the middle classes. Some social scientists privilege income as the central defining variable for class identity, while others make a point of paying attention to net worth and wealth as important variables. However, as my ethnography will show, income alone is not a reliable method of identifying class. Some Brazilian friends who thought of themselves as working class (*classe baixa or classe popular*) had jobs in the formal economy, earned middle-income salaries and were able to make their houses quite comfortable through improvements. Others who classified themselves as middle class (*classe média*) had substantial credit card debt, lived with their parents, had multiple short-term jobs, and took the bus. Indeed, *not* identifying as middle class is valued in some settings, even if social mobility to a higher income neighbourhood has been achieved. Rather than revealing a stable, fixed lifestyle defined by occupation, income, and consumption, the case studies presented in this book reveal people moving unsystematically between different social levels.[23]

[23] An earlier version of these arguments were developed in the author's publication, "Negotiating Inequality: The Contemporary Black Middle Classes in Salvador, Brazil," in *People, Money and Power in the Economic Crisis: Perspectives from the Global South*, edited by Keith Hart and John Sharp (2014, New York and Oxford: Berghahn Books, pp. 106–128).

In this book, I pursue two larger questions related to class: firstly, I am interested in the ways in which class, conceptualized as the general socio-economic structure of Brazil, has undergone several shape changes over time, including periods of enlargement and shrinkage of the middle classes, within a society that has long been unequal. Secondly, I am interested in how people experience class in everyday social interaction, including as an aspect of personal identity. Using my ethnographic data, I will argue that people are never wholly constructed by their class position, or by any other aspect of their identity.

Terms such as "social mobility" and "social stratification" have often been preferred over "class" for describing social inequalities, especially in the United States (Smith 1984, 468). "Social mobility" embodies a belief in the constant movement of individuals into greater prosperity over their life cycle, across generations, or across space unto greater opportunities. However, upward mobility is a kind of double-sided sword; it could mean moving unto better things, but it also involves moving away from what are considered poorer things—poorer people, poorer neighbourhoods, which in Salvador's context is symbolized by blackness. I found it useful to think about these processes through the lens of class and race "projects," a term borrowed from Satre (1963). If class is an object of desire (or repulsion), whether in the past or today, then it seems more useful to think of people, groups, policy makers, culture makers, and so on, as engaged in "class projects" rather than, or in addition to, being occupants of particular classes. These class projects are racialized and always in the making. Thus, people don't just find themselves in a particular social class, they continually create it symbolically, defending or challenging its boundaries, fearing or striving for it.

Intersectionality

Examining the life experiences of Afro-Brazilians at the intersection of race, class, and gender is integral to understanding how social and structural inequality affects their daily lives, including their access to education and job opportunities and their social relationships. Intersectionality emphasizes how social divisions are intermeshed with one another to produce multiple oppressions (Crenshaw 1989). This critical perspective helps to demonstrate the nuanced ways in which, for example, upwardly mobile Afro-Brazilian women faced inequality in their public and personal lives—as well as the different ways in which they responded to this. Black feminist anthropologists have made substantial contributions to research on black women's experiences in Brazil, though the focus has tended to be working-class women (Caldwell 2007; Edo 2016; Falu 2013; Hordge-Freeman 2015; Perry 2013; Williams 2013). My ethnographic material on everyday life in Brazil will reveal how, for black women living and working in upscale areas of the city, practices and discourses surrounding beauty and body presentation were particularly significant and underscored deeper, politicized issues about race, visibility, social mobility, and citizenship in Brazil. These struggles are racialized and gendered in ways that position the black female body as particularly problematic. In Brazil, a dominant discourse on race has valued physical features associated with whiteness, such as having lighter skin, a narrow nose, thin lips, and straight hair. On the other hand, having dark skin, full lips, and a wide nose are associated with African physical features and assigned a low aesthetic value (Edmonds 2010; Caldwell 2007; Gordon 2013; Hordge-Freeman 2015). This highlights special issues for black women, because as a person climbs the social ladder, one is expected to transform one's physical appearance into socially acceptable feminine ideals that emphasize the submersion of African characteristics, although this is changing.[24] Afro-Brazilian women's negotiation of this context of inequality reveals both acceptance and rejection of these dominant ideals and it also reveals often painful experiences of navigating racialized hierarchies.

It is my goal that the material discussed in this book will contribute to greater reflection and critical theoretical analysis of the intersections of race, class, and gender in social life. It is important to note that in this book, I examine the gender norms and identities of cisgender heterosexual women and men—that is, individuals who are labelled male and female at birth and self-identify as men and women. As a result of focusing in-depth on a small number of families and their networks as well as socializing in specific spaces, diversity in gender and sexual identities as presented by this author may appear restricted. I recognize this as a limitation of the research.

[24] Aspects of this argument were developed by the author in a previous publication: "A Beleza abre Portas: Beauty and the Racialized Body among Black Middle-Class Women in Salvador, Brazil," (July 17, 2013, *Feminist Theory* 14: 203–218).

Methodology: Fieldwork with Afro-Brazilian Families

My study took place over 16 months in Salvador, Bahia, Brazil, between 2005 and 2007, a period when the Workers' Party (Partido de Trabalho, or PT) was in power. Follow-up visits in the summers of 2013 and 2017 underscored the value of long-term engagement with Afro-Brazilian families in a specific field-site. This book draws on ethnographic research, observations, and interviews with the members of seven core families who self-identified and/or were identified by others as black middle class, and their extended kin and friendship networks. I carried out semi-formal and unstructured interviews, collecting 115 life histories and genealogies of different generations in the family, whose members also had varied social backgrounds. This provided a sense of how broad historical shifts and cultural changes were experienced by family members of different ages and gender. I participated as much as I could in their daily lives, such as attending family dinners, weddings, birthday parties, socializing in bars and restaurants, going on family vacations, or attending church. For short periods of time, I was able to live with two families in two different neighbourhoods, allowing me to have a more personal view of family dynamics and relationships.

I met respondents through an initial network of friends in Brazil, employing snowball or chain sampling techniques to recruit other respondents. I also used purposive sampling to identify and conduct interviews and focus groups with Black Movement leaders and members. Additionally, I gathered information from local newspapers, research archives, television programmes, and attended popular events (please see Appendix A for a detailed discussion of research methods and research positionality).

The Afro-Brazilian families that are part of this study lived in varied neighbourhoods. Exploring the social networks in which they were embedded revealed the heterogeneity of their lives: family members came from a spectrum of social backgrounds (rural and urban) and had varied career trajectories in the civil service, the professions, education, business, the music industry, Black Movement organizations, banking, food industries, and transportation, to name a few. Some continued to identify strongly with the working-class neighbourhoods they were brought up in and the kin and family networks that they maintained there. However, most people move out of these neighbourhoods as soon as they have the chance, living in heavily guarded, modern apartment buildings or gated communities and socializing in upscale parts of the city where their offices and recreational activities are located. This research dynamic challenged the traditional Western assumption in research of the unequal power relations between interviewer (the person assumed to have more power) and interviewee. Access had to be negotiated and credentials checked: this required that I present official letters, documents, and IDs. Some family members whom I interviewed were not comfortable with being taped or being included in photographs or family pictures. However, I described my research to all interviewees, what I intended to do with the information collected and the precautions I would take to protect their privacy.

Afro-Brazilian Agency and Expanded Citizenship

I discovered that in this type of research, it is critical to learn from the efforts being made by Afro-Brazilians in the struggle against racial oppression and its possible repercussions across the black diaspora. The Afro-Trinidadian historian C.L.R. James wrote *The Black Jacobins* in 1938 because he believed that the history of the most successful slave revolt in the New World, the Haitian Revolution of 1791–1804, was as significant as the American and French Revolutions. James' writing was not simply an exercise in black pride; *The Black Jacobins* ended with reflections on the relevance of the Haitian Revolution for the struggle for African independence from colonial rule. James understood that what happens to black populations in Africa and around the world poses a potentially revolutionary challenge to structures of inequality in the world at large.

In the spirit of C.LR. James, I understand that race as a social principle tends to put people into boxes that do not accurately reflect reality. The people I interviewed and spent time with were not simply victims of racism and inequality, but actively negotiated this system using a mixture of social, cultural, political, and economic strategies. Not all strategies would be regarded as revolutionary from the perspective of black activists or scholars invested in the idea that the black middle classes should be instrumental in changing the conditions of poorer blacks and fighting racial oppression; but there is a desire among many to be considered full citizens, with the right to participate and engage with society, and to change the terms of their positioning in that society and by extension, the world.[25]

[25] These points were explored in the author's previous publication, "Thinking through Fieldwork Encounters in Brazil: Critical Reflections and Future Pathways," in *Decolonizing Qualitative Approaches in the Caribbean*, edited by Saran Stewart (2019, Information Age Publishing Inc.).

Indeed, these are struggles around expanding the notion of citizenship, especially in addressing its class, racial, and gender dimensions. In Brazil and across the black diaspora, the voices of youth and women are rising as they demand a better quality of citizenship—that is, their right not to be treated as second-class citizens. For example, the recent *rolezinhos* (literally, little strolls) that took place in cities such as São Paulo and Rio de Janeiro were mostly organized by poor and black Brazilian youth and those that identified with them. They involved the occupation of malls and other places associated with wealthy, white consumers, thereby challenging racial and economic marginalization and discrimination in Brazil (Vargas 2014). Additionally, black women (especially poor black women) have emerged at the forefront of struggles having to do with land rights, urban housing, socio-spatial exclusion, extractivism, food security, and other issues affecting their communities and livelihoods (Perry 2013; Hernandez-Reyes 2019).[26]

The term "cultural citizenship," first introduced to anthropological audiences by Renato Rosaldo who studied Latino immigrants in the United States, refers to the right to be different (in terms of race, ethnicity, or native language), with respect to the norms of the dominant national community, without compromising one's right to belong, in the sense of participating in the state's political processes (Rosaldo 1994, 57). It is noteworthy that in the aftermath of Marielle Franco's murder in Brazil in 2018, a wave of black women have been inspired to run for office—even as they have encountered glass ceilings in Brazilian politics.[27] In Brazil's current political context and unfolding health crisis, however, there is a need for future studies to look at threats to citizenship rights, especially where the president has openly expressed homophobic, misogynistic, and racist views, as boundaries are drawn around those who belong and those who do not belong.[28]

[26] See the Economic Commission for Latin America and the Caribbean, '*Afrodescendent Women: Debts of Equality*' (Santiago: United Nations, 2018). https://repositorio.cepal.org/bitstream/handle/11362/44387/1/S1800725_en.pdf. Accessed July 23, 2021.

[27] See Gladys Mitchell-Walthour, "'My vote will be Black' – A Wave of Afro-Brazilian Women Ran for Office in 2020 but found Glass Ceiling Hard to Break." *The Conversation*, November 24, 2020. https://theconversation.com/my-vote-will-be-black-a-wave-of-afro-brazilian-women-ran-for-office-in-2020-but-found-glass-ceiling-hard-to-break-150521. Accessed July 21, 2021.

[28] Mariana Simões, 'Brazil's Polarizing New President, Jair Bolsonaro, in His Own Words,' *New York Times*, October 28, 2018. https://www.nytimes.com/2018/10/28/world/americas/brazil-president-jair-bolsonaro-quotes.html. Accessed July 21, 2021.

Seeing your research participants as active agents in changing the condition of their lives is an important part of a critical research perspective. As I will show, Afro-Brazilian scholars and activists are engaged in learning about the history of resistance across the African diaspora and studying and changing the relationship between race, racism, and power. Their activities are key to understanding and challenging processes of racialization and racial discrimination on a wider scale.

I found that the emerging "anthropology of becoming" was a useful lens for studying Afro-Brazilian agency in a highly unequal society such as Salvador (see e.g. Biehl and Locke 2010; Biehl 2017). The anthropology of becoming seeks to remediate Foucauldian and neo-Marxist views of people as "done to" or "done through." It diverts our attention from foreclosures to openings, allowing us to see Afro-Brazilians as not only shaped by larger power structures but also as actors who continually create, reproduce, or ignore dominant discourses and structures. Similar to other process-oriented writings, such as those done under the rubrics of performativity and practice theory, this lens gives primacy to identity as a process of becoming, with an emphasis on desire, focusing on "the ways that desires can open alternative pathways" when individuals and collectives are excluded from preferred pathways (Biehl and Locke 2010, 318). It highlights how Afro-Brazilians might challenge stigmatizing classifications and assumed or imposed limitations. Indeed, dominated groups have rich and complex cultural, political, and religious lives that support diverse oppositional activities.

Organization of This Book

The first two chapters (Chaps. 2 and 3) of this book explore processes of racialization by focusing on the larger context of inequality. I therefore lay the groundwork for understanding the lives and experiences of my research participants and consider the impact of broader patterns of social change and transformations on the lives of Afro-Brazilians. Thus, there is a sense in which "Big History" (the history of social transformations) is significant here. I do not necessarily subscribe to the idea that there have been massive historical and dramatic shifts from colonial Brazil to the present time: nevertheless, there are clear changes, especially from the 1950s to the present. These changes were brought about by human action and so my concern is to show the interplay between social transformations, state discourses and actions, and people's lives. I will explore different racial discourses in Brazil and describe participants' experiences with racism and their relationships to racial identity.

In Chaps. 4 and 5, I examine the class biographies of research participants within the context of a racialized society, giving insight into varied strategies of social mobility. I will attempt to map the contours of the contemporary black middle classes in Salvador, underscoring the heterogeneity and diversity in their life trajectories. In general, I explore the relational, symbolic, and material practices that research participants utilized, including practices of distinction between black middle classes and black lower classes. In Chap. 6, I pay particular attention to family and kinship networks, thereby addressing a tendency in the classic literature on black social mobility in Brazil to privilege the individual in social ascension and overlook the social context in which he or she is embedded. Chapter 7 will explore practices and discourses focused on the body and physical appearance, including the importance of *boa aparência* (good appearance) in social assessments of class positioning and as an avenue of social mobility.

I advance three main arguments in this book. Firstly, while race and class are inextricably intertwined in Salvador's hierarchies, I argue that they do not necessarily map unto each other in straightforward and automatic ways. Symbolic practices of "whitening" and "blackening" can be complex and contradictory processes. Furthermore, these processes highlight the contentious issue of the relationship between the upwardly mobile black person and his/her former community which can entail racialized judgements of moral personhood and character. Secondly, I argue that Afro-Brazilian middle-class identities are diverse and varied and that different modes of ascension into the middle classes exist. Of course, this complicates the meaning of the Afro-Brazilian "community." The increased promotion and appropriation of Afro-Brazilian (and especially Afro-Bahian) culture by a plurality of social actors such as the state, tourism agencies, intellectuals, artists, musicians, and activists imply that there is a stable, continuous, and well-bounded set of traditions that are shared by all people of African descent in Brazil. I focus instead on competing representations of constructing "community" and what it means to be Afro-Brazilian, revealing multiple race and class "projects" rather than one single, unifying one. Finally, I argue that Afro-Brazilians are engaged in expanding notions of citizenship and belonging in Brazil through different spheres of life—such as their involvement in state-led and other social programmes, Afro-Brazilian organizations, religious groups, and through their beauty practices and discourses. Some of these strategies (such as beauty practices) may not be interpreted as radical challenges to the race/class structure. However, I show that they have potential revolutionary effect.

References

Alcoff, L.M. 2006. *Visible Identities: Race, Gender, and the Self.* Oxford/New York: Oxford University Press.

Alcoff, L.M., M. Hames-Garcia, S.P. Mohanty, and P.M.L. Moya. 2006. *Identity Politics Reconsidered (Future of Minority Studies).* New York: Palgrave MacMillan.

Andrews, G.R. 1991. *Blacks and Whites in São Paulo, Brazil. 1888–1988.* Madison/Wisconsin: University of Wisconsin Press.

Anthias, F., and Nira Yuval-Davis. 1992. *Racialized Boundaries: Race, Nation, Gender, Class and the Anti-Racist Struggle.* London: Routledge.

Bailey, S. 2009. *Legacies of Race: Identities, Attitudes, and Politics in Brazil.* Stanford, CA: Stanford University Press.

Banton, M. 2015. *What We Now Know About Race and Ethnicity.* New York and Oxford: Berghahn.

Barbosa, L. 1992. *Jeitinho: A Arte de Ser Mais Igual do que os Outros.* Rio de Janeiro: Editora Campus.

Beckett, G. 2017. The Abolition of All Privilege: Race, Equality, and Freedom in the Work of Antenor Firmin. *Critique of Anthropology* 37 (2): 160–178.

Biehl, J., and Peter Locke. 2010. Deleuze and the Anthropology of Becoming. *Current Anthropology* 51 (3): 317–351.

Biehl, J., ed. 2017. *Unfinished: The Anthropology of Becoming.* Durham: Duke University Press.

Bonilla-Silva, E. 2018. *Racism Without Racists: Color-Blind Racism and the Persistence of Racial Inequality in America.* Lanham and London: Rowman and Littlefield Publishers.

Bourdieu, P. 1984. *Distinction: A Social Critique of the Judgement of Taste.* Cambridge, MA: Harvard University Press.

Bowser, B. 2007. *The Black Middle Class: Social Mobility—And Vulnerability.* Boulder and London: Lynne Reinner Publishers.

Burdick, J. 2005. Why Is the Black Evangelical Movement Growing in Brazil? *Journal of Latin American Studies* 37: 311–332.

Caldwell, K.L. 2007. *Negras in Brazil: Re-envisioning Black Women, Citizenship, and the Politics of Identity.* New Brunswick: Rutgers University Press.

Campante, F.R., Anna R.V. Crespo, G.P.G. Phillippe, and Leite. 2004. Desigualdade Salarial Entre Raças no Mercado de Trabalho Urbano Brasileiro: Aspectos Regionais. *Revista Brasileira de Economia* 58 (2): 185–210.

Castro, N.A., and A.S. Guimarães. 1993. Desigualdades Raciais no Mercado e nos Locais de Trabalho. *Estudos Afro-Asiáticos* 24: 23–60.

———, N.A., and V.S. Barreto. 1998. *Trabalho e Desigualdade Raciais: Negros e Brancos no Mercado de Trabalho em Salvador*. São Paulo: Annablume.

Ciconello, A. 2008. The Challenge of Eliminating Racism in Brazil: The New Institutional Framework for Fighting Racial Inequality. In *From Poverty to Power: How Active Citizens and Effective States Can Change the World*, ed. Mark Fried and Duncan Green, 1–14. Oxford: Oxfam International. https://oxfamilibrary.openrepository.com/bitstream/handle/10546/112382/fp2p-cs-challenge-elminating-racism-brazil-140608-en.pdf;jsessionid=F5CE7B8CB3FC9F3D1E01BA2DBF6CADC9?sequence=1. Accessed 23 July 2021.

Cohen, S. 2004. *Searching for a Different Future: The Rise of a Global Middle Class in Morocco*. Durham/London: Duke University Press.

Comaroff, J. 1987. Of Totemism and Ethnicity: Consciousness. *Practice and Signs of Inequality. Ethnos* 52 (3–4): 301–323.

Costa-Pinto, Luiz. (1953) 1998. *O Negro no Rio de Janeiro: Relações de Raça Numa Sociedade em Mudança*. Rio de Janeiro: UFRJ.

Crenshaw, K. 1989. Demarginalizing the Intersection of Race and Sex: A Black Feminist Critique of Antidiscrimination Doctrone, Feminist Theory, and Antiracist Politics. University of Chicago Legal Forum, 1989, Article 8. https://chicagounbound.uchicago.edu/uclf/vol1989/iss1/8/. Accessed 26 July 26, 2021.

Da Costa, A.E. 2014. *Reimagining Black Difference and Politics in Brazil: From Racial Democracy to Multiculturalism*. New York: Palgrave Macmillan.

Da Silva Soares, R. 2004. Negros de Classe Média em São Paulo: Estilo de Vida e Identidade Negra. PhD dissertation, Universidade de São Paulo, .

De Santana, I. 1999. Executivos Negros em Organização Bancários de Salvador: Dramas e Tramas do Processo de Ascensão Social. Revista Afro-Ásia 23: 199:240. Salvador: CEAO/UFBA.

———. 2009. À Margem do Centro: Ascensão Social e Processos Identitários Entre Negros de Alto Escalão: O Caso de Salvador. PhD dissertation, Universidade Federal da Bahia.

Degler, C. 1971. *Neither Black nor White: Slavery and Race Relations in the United States and Brazil*. New York: Macmillan.

Do Valle Silva, N. 1985. Updating the Cost of Not Being White in Brazil. In *Race, Class and Power in Brazil*, ed. Pierre Michel Fontaine, 42–55. Los Angeles: UCLA Center for Afro-American Studies.

Dos Santos, S.B. 2008. Brazilian Black Women's NGOs and Their Struggles in the Area of Sexual and Reproductive Health: Experiences, Resistance, and Politics. PhD dissertation, University of Texas at Austin,

DuBois, W.E.B. 1899. *The Philadelphia Negro: A Social Study*. Philadelphia: University of Philadelphia Press.

———. [1903] 1990. *The souls of Black folk*. New York: Vintage Books.

ECLAC. 2018. Afro-descendent women in Latin America and the Caribbean: Debts of equality. Economic Commission for Latin America and the Caribbean (in cooperation with Cooperación Espanola). Santiago: United Nations. https://repositorio.cepal.org/bitstream/handle/11362/44387/1/S1800725_en.pdf. Accessed 23 July 23, 2021.

Edmonds, A. 2010. *Pretty Modern: Beauty, Sex, and Plastic Surgery in Brazil*. Durham and London: Duke University Press.

Edo, Ugo. 2016. Beauty and the Black: Fertility Control in an Economy of Aesthetics, Race, and Sexuality in Brazil. Paper Presented at the Annual Meeting of the American Anthropological Association, Minneapolis, November.

Falu, Nessette. 2013. Invisible Hands: An Epistemology of Black Religious Thought and Black Lesbian Sexual Desire that Disrupts Crystallized Culture. In *Ain't I a Womanist Too: Third Wave Womanist Religious Thought*, ed. Monica A. Coleman, 149–160. Minneapolis: Fortress Press.

Fernandes, F. 1969. *The Negro in Brazilian Society*. New York: Columbia University Press.

Figueiredo, A. 2002. *Novas Elites de Cor: Estudos Sobre os Profissionais Liberais Negros de Salvador*. São Paulo: Annablume.

———. 2003. A Classe Media não vai ao Paraíso: Trajetórias, Perfis e Negritude entre Os Empresarios Negros. PhD dissertation, Universidade do Rio de Janeiro.

———. 2004. A Classe Media não vai ao Paraíso: Trajetórias, Perfis e Negritude entre Os Empresarios Negros. PhD dissertation, Universidade do Rio de Janeiro, 2004. Fora do Jogo: A Experiência dos Negros na Classe Media Brasileira. Cadernos Pagu 23: 199–228.

Fontaine, P.M. 1980. Research in the Political Economy of Afro-Latin America. *Latin American Research Review* 15 (2): 111–141.

Frankenburg, R. 1993. *White Women, Race Matters: The Social Construction of Whiteness*. Minneapolis: University of Minnesota Press.

Frazier, E.F. 1957. *Black Bourgeoisie: The Rise of a New Middle Class*. New York: Free Press.

Freyre, G. 1956. *The Masters and the Slaves (Casa-Grande e Senzala): A Study in the Development of Brazilian Civilization*. London: Weidenfeld and Nicolson.

———. 1961. O Luso e o Trópico; Sugestões em Turno dos Métodos Portugueses de Integração de Povos Autóctones e de Culturas Diferentes da Européia num Complex Novo de Civilização: O Luso-Tropical. Lisboa: Comissão Executiva das Comemorações do V Centenário da Morte do Infante D. Henrique.

Fry, P. 1995–96. O Que a Cinderela Negra tem a Dizer sobre a 'Política Racial' no Brasil. Revista da USP 28: 122–135.

Garvey, A.J., ed. 1967. *The Philosophy and Opinions of Marcus Garvey or Africa for the Africans*. London: Frank Cass and Company Ltd.

Giacomini, S.M. 2006. *A Alma da Festa: Família, Etnicidade e Projetos num Clube Social da Zona Norte do Rio de Janeiro: O Renascenca Clube*. Belo Horizonte: Universidade Federal de Minas Gerais.

Gilroy, P. 1993. *The Black Atlantic: Modernity and Double Consciousness*. London: Verso.

Golash-Boza, T. 2010. 'Had They Been Polite and Civilized, None of This Would Have Happened': Discourses of Race and Racism in Multicultural Lima. *Latin American and Caribbean Ethnic Studies* 5 (3): 317–330.

Goldberg, D.T. 1993. *Racist Culture: Philosophy and the Politics of Meaning*. Oxford: Blackwell.

———. 2008. *The Threat of Race: Reflections on Racial Neoliberalism*. Malden, MA: Wiley-Blackwell.

Gomes, D. 2019. Brazilian Politics and the Rise of the Far Right. *Black Perspectives*, 7 January. https://www.aaihs.org/brazilian-politics-and-the-rise-of-the-far-right/. Accessed 24 July 2021

Gordon, D. 2013. A Beleza Abre Portas: Beauty and the Racialized Body Among Black Middle-Class Women in Salvador, Brazil. *Feminist Theory* 14 (2): 203–218.

Graham, L.O. 1999. *Our Kind of People: Inside America's Black Upper Class*. New York: Harper Collins Publishers.

Guimarães, A.S.A. 2001. Brazil: The Misadventures of Nonracialism in Brazil. In *Beyond Racism: Race and Inequality in Brazil, South Africa, and the United States*, ed. Charles Hamilton, Lynn Huntley, Neville Alexander, Antonio Sérgio Alfredo Guimarães, and Wilmot James. Boulder: Lynne Reinner Publishers.

———. 2006. Depois da Democracia Racial. *Tempo Social, Revista de Sociologia da USP* 18 (2): 269–287.

Hall, S. 1997. The Problem of Ideology: Marxism Without Guarantees. In *Stuart Hall: Critical Dialogues in Cultural Studies*, ed. D. Morley and K.H. Chen. New York: Routledge.

Hasenbalg, C.A. 1979. *Discriminação e Desigualdades Raciais no Brasil*. Rio de Janeiro: Graal.

———. 1985. Race and Socioeconomic Inequalities in Brazil. In *Race, Class and Power in Brazil*, ed. P.M. Fontaine. Los Angeles: Centre for Afro-American Studies, UCLA.

Hasenbalg, C.A., and N. do Valle Silva. 1991. Raça e Oportunidades Educacionais no Brasil. In *Desigualdade Racial no Brasil Contemporâneo*, ed. Peggy A. Lovell. Belo Horizonte: UFMG.

Hernandez-Reyes, C.E. 2019. Black Women's Struggles Against Extractivism, Land Dispossession, and Marginalization in Colombia. *Latin American Perspectives* 46 (2): 217–234.

Hordge-Freeman, E. 2015. *The Colour of Love: Racial Features, Stigma and Socialization in Black Brazilian Families*. Austin: University of Texas Press.

James, C.L.R. 1838. *The Black Jacobins: Toussaint Louverture and the San Domingo Revolution*. London: Secker and Warburg.

Kottak, C.P. 2015. *Cultural Anthropology: Appreciating Cultural Diversity.* New York: McGraw Hill Education.

Landry, B. 1987. *The New Black Middle Class.* Oakland, CA: University of California Press.

Lima, M. 2001. Serviço de 'Branco,' Serviço de 'Preto': um Estudo sobre Cor e Trabalho no Brasil Urbano. PhD Dissertation, Universidade Federal do Rio de Janeiro.

Lovell, P.A. 1999. Women and Racial Inequality at Work in Brazil. In *Racial Politics in Contemporary Brazil*, ed. Michael Hanchard. Durham/London: Duke University Press.

Mignolio, W.D. 2011. *The Darker Side of Western Modernity: Global Futures, Decolonial Options.* Durham: Duke University Press.

Miles, R., and M. Brown. 2003. *Racism.* London/New York: Routledge.

Mitchell-Walthour, G. 2020. My Vote Will Be Black: A Wave of Afro-Brazilian Women Ran for Office in 2020 But Found the Glass Ceiling Hard to Break. *The Conversation*, November 24. 'My vote will be Black' – A Wave of Afro-Brazilian Women Ran for Office in 2020 But Found Glass Ceiling Hard to Break (theconversation.com). Accessed 21 July 2021.

———. 2017. Economic Pessimism and Racial Discrimination in Brazil. *Journal of Black Studies* 48 (7): 675–697.

Neri, M.C. 2008. The New Middle Class. Rio de Janeiro: Fundação Getúlio Vargas/Centro de Políticas Sociais. https://www.cps.fgv.br/ibrecps/M3/M3_MidClassBrazil_FGV_eng.pdf. Accessed 26 July 26, 2021.

Nogueira, O. 1985. *Tanto Preto Quanto Branco: Estudos de Relações Raciais.* São Paulo: T.A. Queiroz.

O'Dougherty, M. 2002. *Consumption Intensified: The Politics of Middle Class Daily Life in Brazil.* Durham and London: Duke University Press.

Omi, M., and Howard Winant. 1994. *Racial Conditions: Politics, Theory, Comparisons.* Minneapolis: University of Minnesota Press.

———. 2014 [1986]. *Racial Formation in the United States.* New York: Routledge.

Ortner, S. 2003. *New Jersey Dreaming: Capital, Culture, and the Class of '58.* Durham and London: Duke University Press.

Owensby, B.P. 1999. *Intimate Ironies: Modernity and the Making of Middle Class Lives in Brazil.* Stanford: Stanford University Press.

Paixão, M. 2004. Waiting for the Sun: An Account of the (Precarious) Social Situation of the African Descendant Population in Contemporary Brazil. *Journal of Black Studies* 34 (6): 743–765.

Parker, J. 2009. Burgeoning Bourgeoisie. *The Economist* (Special Report), 14 February. https://www.economist.com/special-report/2009/02/14/burgeoning-bourgeoisie. Accessed July 23, 2021.

Patillo-McCoy, M. 1999. *Black Picket Fences: Privilege and Peril among the Black Middle Class.* Chicago/London: The University of Chicago Press.

Perlman, J. 2004. The Metamorphosis of Marginality in Rio de Janeiro. *Latin American Research Review* 39 (1): 189–192.

Perry, K.K.Y. 2013. *Black Women Against the Land Grab: The Fight for Racial Justice in Brazil.* Minneapolis: University of Minnesota Press.

Pucci, Claudio. 2016. *Social, Racial and Gender Profile of the 500 Largest Brazilian Companies.* São Paulo: Instituto Ethos. https://publications.iadb.org/en/social-racial-and-gender-profile 500-largest-brazilian-companies. Accessed September 27, 2020.

Quijano, A. 2007. Coloniality and Modernity/Rationality. *Cultural Studies* 21 (2–3): 168–178.

Ramos, G.A. 1995. *Introdução Crítica a Sociologia Brasileira.* Rio de Janeiro: Universidade Federal do Rio de Janeiro.

Rede de Observatórios de Segurança. 2020. A Cor de Violência na Bahia – Uma Análise dos Homicidios e Violência Sexual na Ultima Década, 4–6. http://observatorioseguranca.com.br/wp-content/uploads/2020/03/A-cor-da-viole%CC%82ncia-na-Bahia-Uma-ana%CC%81lise-dos-homici%CC%81dios-e-viole%CC%82ncia-sexual-na-u%CC%81ltima-de%CC%81cada-FINAL.pdf. Accessed July 23, 2021.

Restrepo, E., and A. Rojas. 2010. *Inflexión Decolonial: Fuentes, Conceptos y Cuestionamientos.* Popayan: Editorial Universidad del Cauca.

Rodney, W. [1973] 1983. *How Europe Underdeveloped Africa.* London and Tanzanian Publishing House, Dar-Es-Salam: Bogle-L'Ouverture Publications.

Rosaldo, R. 1994. Cultural Citizenship and Educational Democracy. *Cultural Anthropology* 9 (3): 402–411.

Roth-Gordon, J. 2017. *Race and the Brazilian Body: Blackness, Whiteness, and Everyday Language in Rio de Janeiro.* California: University of California Press.

Ryan, W. 1976. *Blaming the Victim.* New York: Vintage.

Sansone, L. 2003. *Blackness Without Ethnicity: Constructing Race in Brazil.* New York: Palgrave Macmillan.

Satre, J.P. 1963. *Search for a Method.* New York: Alfred A. Knopf.

Silva, G.M.D., and E.P. Reis. 2011. Perceptions of Social Mobility, Inequality and Racial Discrimination among Black Professionals in Rio de Janeiro. *Latin American Research Review* 46: 55–78.

———. 2012. The Multiple Dimensions of Racial Mixture in Rio de Janeiro, Brazil: From Whitening to Brazilian Negritude. *Ethnic and Racial Studies* 35 (3): 382–399.

Simões, M. 2018. Brazil's Polarizing New President, Jair Bolsonaro, in His Own Words. *New York Times*, October 28. https://www.nytimes.com/2018/10/28/world/americas/brazil-president-jair-bolsonaro-quotes.html. Accessed July, 23, 2021.

Skeggs, B. 1997. *Formations of Class and Gender.* London: Sage Publications Ltd.

Smith, C.A. 2016. *Afro-Paradise: Blackness, Violence, and Performance in Brazil.* Urbana/Springfield: University of Illinois Press.

Smith, R.T. 1984. Anthropology and the Concept of Social Class. *Annual Review of Anthropology* 13: 467–494.

Stolcke, V. 1995. Talking Culture: New Boundaries, New Rhetorics of Exclusion in Europe. *Current Anthropology* 36 (1): 1–23.

Sue, C.A. 2013. *Land of The Cosmic Race: Race Mixture, Racism and Blackness in Mexico.* New York: Oxford University Press.

Telles, E.E. 2004. *Race in Another America: The Significance of Skin Colour in Brazil.* Princeton: Princeton University Press.

Thomas, D.A., and M. Kamari Clarke. 2013. Globalization and Race: Structures of Inequality, New Sovereignties, and Citizenship in a Neoliberal Era. *Annual Review of Anthropology* 42: 305–325.

Turra, C., and G. Venturi. 1995. *Racismo Cordial: A Mais Completa Análise sobre o Preconceito de Cor no Brasil.* São Paulo: Editora Ática.

Twine, F.W. 1998. *Racism in a Racial Democracy: The Maintenance of White Supremacy in Brazil.* New Brunswick: Rutgers University Press.

Vargas, J.H.C. 2014. Black Disidentification: The 2013 Protests, Rolezinhos, and Racial Antagonism in Post-Lula Brazil. *Critical Sociology* 42 (4–5): 551–565.

Wade, P. 2001. Racial Identity and Nationalism: A Theoretical View from Latin America. *Ethnic and Racial Studies* 24 (5): 845–865.

———. 2005. Rethinking Mestizaje: Ideology and Lived Experience. *Journal of Latin American Studies* 37: 239–257.

———. [1997] 2010. Race and Ethnicity in Latin America. London/New York: Pluto Press.

Wade, P., J. Scorer, and I. Aguiló, eds. 2019. *Cultures of Anti-Racism in Latin America and the Caribbean.* London: University of London.

Williams, E. 1994. *Capitalism and Slavery.* Chapel Hill/London: University of North Carolina Press.

Williams, E.L. 2013. *Sex Tourism in Bahia: Ambiguous Entanglements.* Champaign: University of Illinois Press.

CHAPTER 2

The Context of Inequality

To fully contemplate the lives of upwardly mobile African descendants in contemporary Brazil, it is necessary to examine black social mobility within the context of the country's historical, social, and political-economic landscape. While Brazil is a significant economic player in Latin America, it is also a country of dramatic socioeconomic inequalities, rooted in its colonial history and political development. At the start of doctoral fieldwork in 2005, income inequality in Brazil was one of the highest in the world, with the richest 20 per cent of the population earning 33 times more than the 20 per cent of the population living on the lowest income (World Bank 2004).[1] More recently, the income gap between the country's top and bottom decile remains about five times as wide as in advanced nations.[2] Traditionally, these social inequalities have been explained in terms of class in Brazil, in a country where racial mixture was seen as crucial to national identity, as well as the reason for low racism in the country.[3] Most contemporary scholars agree that this image of Brazil as a racial democracy in the early twentieth century was overstated, while a number of studies

[1]World Bank. 2004. *Inequality and Economic Development in Brazil.* https://elibrary.worldbank.org/doi/abs/10.1596/0-8213-5880-4. Accessed July 30, 2021.

[2]See World Economic Forum news bulletin by Gemma Corrigan, 7 September 2015, *Is Brazil Making Progress on Inequality?* https://www.weforum.org/agenda/2015/09/is-brazil-making-progress-on-inequality/. Accessed July 30, 2021.

[3]In the United States, unlike Brazil, race (rather than class) is commonly presented as the more powerful cause of social divisions.

D. J. Gordon, *Blackness and Social Mobility in Brazil*,
https://doi.org/10.1007/978-3-030-90765-5_2

have shown that social gaps are racialized and gendered in significant ways and cannot be explained solely in terms of class.[4] For example, Table 2.1 shows differences in social indicators between whites and non-whites (black and brown Brazilians) for the year 1996–1997 across measures such as the Human Development Index, life expectancy, adult literacy, and school enrolment, indicating racial inequalities.

Ideas about race and difference have profoundly shaped social inequalities in Brazil, dating back to colonial times. Brazil was the largest slave society in the Americas, and this had a profound effect on the country's social relations, cultural traditions, and family relationships (Hita 2014). Indeed, Brazil retains the notorious reputation of being the last country in the Western Hemisphere to abolish slavery, which occurred at least in part because of concerns over its international image. Even after slavery ended, this did not end racial domination—a pervasive correlation between racism and poverty continued (Hordge-Freeman 2015; Hita 2014). As we will see in the following paragraphs, Afro-Brazilians managed to find ways to negotiate unequal society—including utilizing traditional patterns of social mobility, taking advantage of historical ruptures and transformations, or going into new and expanding areas of work (Guimarães et al. 1995; De Azevedo 1996; Bacelar 1999, 2001; Figueiredo 2003).

Table 2.1 Human Development Index, 1996–1997, indicating racial inequalities in Brazil

	Human Development Index	*Life expectancy at birth (1997)*	*Infant mortality/1000 (1996)*	*Adult literacy rate (1997)*	*School enrolment ratio (1997)*
Brazil	0.773	67	48	85	78
Whites	0.784	70	37	92	82
Non-whites	0.663	64	62	72	73

Source: Edward Telles (2004, 111)

[4] See classic studies such as Fernandes (1969), Hasenbalg (1979), Do Valle Silva (1985), Nogueira (1985), and Costa-Pinto ([1953] 1998).

Salvador, Bahia, Brazil

The beautiful and historic city of Salvador and the state of Bahia are the focus of this book, although commonalities can be found with other Brazilian cities. Founded in 1549 by the Portuguese colonialist Tomé de Souza, Salvador was Brazil's first colonial capital, overlooking the magnificent Baía de Todos os Santos (Bay of All Saints). Historically, Salvador was a wealthy city that rose to national and regional pre-eminence due to the riches gained from sugar, commerce, and the slave trade, connecting it to the entire Portuguese Empire. Beginning in the sixteenth century, roughly 4 million enslaved Africans—or 40 per cent of all enslaved Africans in the Americas—entered Brazil to work on enormous *fazendas* (plantations) of the Northeast, including the fertile Recôncavo area of Bahia (Graden 2006). This shaped the demography, society, and culture of the area and paved the way for the city's impressive monikers—such as "Roma Negra" (Black Rome) of the Americas. Indeed, the city (and Bahia more generally) has strong connections to West African culture through cuisine, music, religion, and language. Salvador is well known throughout the region for its vibrant black cultural life, religious festivals, artistic performances, and open-air Capoeira demonstrations (a type of martial art that originated in Angola). Thus, it is considered a mecca for Africans in the diaspora, attracting thousands of tourists and travellers each year seeking to explore Afro-diasporic identity, religion, music, politics, sexuality, and other cultural activities. As several scholars have argued, Bahia's vibrant and dynamic black culture has both local and transnational dimensions, given that it has emerged as a significant locale in the Atlantic world for African-referenced identities and black cultural and social formations (Perry 2013; De Santana Pinho 2008, 2010; Selka 2007). Map 2.1 is a country map of Brazil indicating where the Northeastern state of Bahia is located while Map 2.2 shows the city of Salvador.

Due to the large African-descended population and its African influenced culture and traditions, racial discourse has long taken on magnified proportions in Bahia. For example, while Bahia has been alternatively romanticized as the "cradle of Brazilian national identity" and as the "Black Rome of Brazil," it has also been denigrated as an embarrassing symbol of Brazil's backwardness. Two equally powerful representations of Bahia emerged in the literature and cultural productions about the city and remain important in the popular imagination. On the one hand, Salvador is seen as a site of mythical origin, where races and cultures mixed and blended, nurturing the birth of Brazilian civilization ("Brown Bahia"). The city was often referred

Map 2.1 Map of Brazil

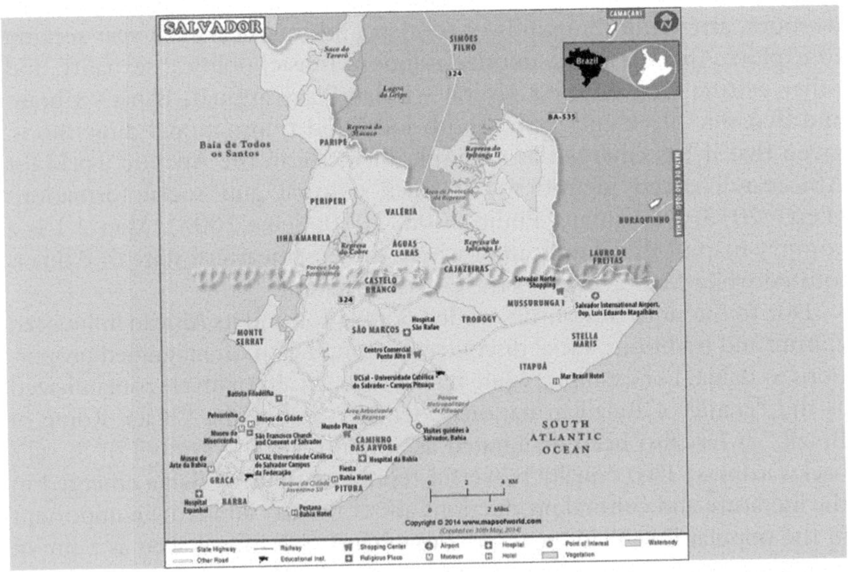

Map 2.2 Map of Salvador

to as "A Mulata Velha," or the "Old Black Mammy," a term that draws on the symbolism of plantation society, conjuring up the image of the generous, loyal, caring wet nurse or nanny that nurtured the children of the colonial white Brazilian family (Otovo 2016, 7). On the other hand, Salvador has the reputation as the "African" capital of Brazil, reflected in its people, language, culture, food, and religion ("Black Bahia"). Black women were particularly associated with how the city was imagined. Highly visible in public life, black women could be seen working as household servants or washerwomen, selling street foods or selling in the market, and walking around as mothers with their children (Otovo 2016, 8). Figure 2.1 is an image of a black Bahian woman taken around the late nineteenth century by the photographer Marc Ferrez (1843–1923).

Fig. 2.1 Portrait of a woman, Bahia, Brazil. (Source: From the New York Public Library https://digitalcollections.nypl.org/items/510d47d9-3c7c-a3d9-e040-e00a18064a99)

American anthropologist Ruth Landes focused on these gendered dynamics and gendered forms of power in her book, *City of Women*, about the mother-priestesses (*mães de santos*) of the Afro-Brazilian religion, Candomblé, in Salvador (Landes 1947). Candomblé is a religion that has roots in the religious traditions of West and Central Africa, incubated in Brazil against the backdrop of slavery, colonialism, and mercantilism (Harding 2000). Maria Gabriella Hita, a sociologist at the University of Bahia, has written a rich ethnography on black matriarchal families in Salvador in the 1990s in her book *A Casa das Mulheres N'Outro Terreiro* (Hita 2014). In this book, she underscores the importance of gender, expansive kinship relations (including spiritual kinship), and territory within Afro-Brazilian *terrieros* (Candomblé temples or places of worship where some adherents live). Thus, the image of the strong, black woman or *mãe preta* (Black Mother) of Candomblé—a spiritual, maternal figure who is everyone's mother—has long been associated with Bahia.

However, this African-derived heritage has not always been accepted. For example, Dixon (2016, 14) has argued that historically and to some extent today, one of the most significant barriers to social standing in Salvador is the open embrace of African cultural practices. Long regarded as dangerous and a form of counter-modernity, Bahian elites traditionally regarded African cultural traditions in a negative light until around the 1930s, when new forces at play radically repositioned the value of Afro-Bahian cultural practices and contributions (Ickes 2013; Romo 2010).[5] This point is further discussed in Chap. 3. For now, it is important to note that these racialized and gendered representations of "Brown Bahia" and "Black Bahia" draw on parallel models of race: namely *mestiçagem* (hybridity) and ethnic absolutism (purity), which have long been part of the Bahian social landscape. It is important to locate Bahia within these two modes of representation.

The modern city of Salvador is the third most populous city in Brazil and is among the 120 largest cities of the world. According to the 2010 census, Salvador is 51.7 per cent *pardo* (brown or mixed race), 28 per cent

[5] Factors contributing to the uptake of Afro-Bahian culture by political and social elites include the politics of Getúlio Vargas (1930–1945), whose policies promoted Afro-Brazilian—and especially Afro-Bahian culture; the writings of Brazilian academics and foreign intellectuals, such as the UNESCO studies of the 1950s; and the role played by Candomblé leaders in reasserting their cultural agency.

preto (black), and 18.9 per cent *branco* (white).[6] There are roughly 3 million residents living in the city proper, with another 3.5 million people living in the surrounding area or periphery of the city (IBGE 2010).[7] Bahia is well known globally for its lively tourism industry centred on the beautiful coast and the marketing of Afro-Brazilian culture such as music (including Afro-rooted art forms such as Reggae and Hip-Hop), cuisine, Capoeira, Candomblé, theatre, and dance.

Given Salvador's cultural and religious importance for Afro-Brazilian and Afro-diasporic culture, one would assume that conditions would be more favourable than most places in present-day Brazil to dark-skinned people—yet the position of black people in Salvador remains ambiguous. Race and class inequality have long been described as extreme and entrenched (McCallum 2005). Most of the city's residents are poor and visibly darker-skinned. In addition, Afro-descendant women and children predominate among the very poorest of the city's residents (McCallum 2007). Afro-Bahian residents also confront considerable structural disadvantages in comparison to the Euro-Bahian elite—such as income disparity, police brutality, high levels of violence that plague the city, and substandard and underfunded public schools that make it difficult for them to achieve mobility (Hordge-Freeman 2015; Perry 2013; Smith 2016). Figure 2.2 is a photograph of a public school taken by the author in the peripheral neighbourhood of Paripe, where majority of the residents are black and brown Brazilians.

Salvador's social inequalities and contradictions are reflected in the spatial organization of the city. One of the more obvious features of the city's topography (though not necessarily most important) is its division into the Higher City (Cidade Alta) and the Lower City (Cidade Baixa). Cidade Alta, built on the plateaus of the city, has long been associated with the more privileged classes where the quieter, cleaner, and more peaceful administrative and residential areas are located, including the first medical school in Latin America. It is linked to the Lower City by steep roads, pathways, trolley cars, and elevators. Cidade Baixa is associated with the

[6] Other categories included in censuses are *amarelo* (yellow, referring to those of Japanese or Chinese origin) and *indígena* (Native Indian or indigenous). However, according to the 2010 censuses, these categories represent about 2 per cent of Brazil's overall population, so they are not included here.

[7] Institute of Brazilian Geography and Statistics figures accessed August 5, 2021. https://cidades.ibge.gov.br/brasil/ba/salvador/panorama

Fig. 2.2 Municipal school, Paripe, Salvador. (Source: Doreen Gordon)

old commercial district and port and was the centre of military defence in colonial times. It has a poorer and darker-skinned residential population, though some middle-class neighbourhoods such as Ribeira and Monte Serrat are located here. Figure 2.3 illustrates the division between the Upper and Lower City in Salvador and Fig. 2.4 shows the seaside area of Ribeira in the Lower City.

By the early twentieth century, other spatial inequalities began to take shape. The Euro-Bahian elite had moved away from their *sobrados* (mansions) and *solares* (colonial buildings with courtyards) in the historic city centre—Pelourinho—where they had once shared space with the black population in a more intimate way. They fled to places like Vitoria, where they built mansions and gardens (Borges 1992). Further up the coast from Vitoria are Barra and Rio Vermelho, whose residents traditionally depended on fishing, subsistence farming, and some handicrafts. Today these areas are associated with the Orla or Avenida Atlântica which runs along the west side of the Bay of All Saints and is known as having some

Fig. 2.3 Lower and Upper City, Salvador. (Source: Doreen Gordon)

of the city's *bairros finos* (fine neighbourhoods) and *gente bonita* (beautiful people).[8]

By the late nineteenth and early twentieth centuries, an early industrialization began to take place which mainly targeted a regional market. Working-class neighbourhoods developed around factories—but this early industrialization was more limited than what took place in other Brazilian cities (Souza 2000). Since the 1940s, there has also been a marked increase in the occupation of the periphery of the city, originating in intensified waves of low-income migrants due to the lack of work in Northeastern towns and cities. This devastating "peripheralization" of the city has been distinctly associated with black bodies.

[8] The phrase *gente bonita* literally translates into "beautiful people." While beauty is generally seen as a quality that can cross lines of class and colour, this term also suggests a divided Brazilian society. *Gente bonita* typically refers to people who live in the wealthier and more "refined" areas of the city, who drive around in cars and who conform to dominant cultural norms that value Euro-Brazilian aesthetics and lifestyles. *Gente bonita* could range from lighter skinned to darker skinned people, but it was not typically associated with places imagined as racially segregated, such as lower-class suburbs.

Fig. 2.4 Ribeira, Lower City. (Source: Doreen Gordon)

In more recent years, *invasões* (invasions), as the illegal occupation of land is sometimes called, have occurred all over the city, such as in areas surrounding middle- and high-income neighbourhoods as well as along the main Salvador-Rio highway and the Avenida Paralela. While these spatial geographies are racialized, it is also important to note that wealth has long been a significant factor in Salvador's spatial distribution (Butler 1998, 135). While my research participants' lives were shaped by these local hierarchies and inequalities, it is also important to have a sense of the larger historical politico-economic and socio-cultural transformations in order to avoid falling into the trap of essentialism that already plagues cultural understandings of Salvador and Bahia more generally.

Afro-Brazilian Social Mobility in Perspective

A new generation of scholars writing on Brazilian social relations have increasingly emphasized the overriding violence imposed on black bodies which characterized the black presence in the Americas (such as forced enslavement; colonialism; the physical and sexual subjugation of black women; death as a reality and an ever present possibility).[9] As Brazil transitioned from a monarchy into a republic, maintaining social order signified control of the "dangerous classes" or disenfranchised masses, so that elites could enjoy the benefits of progress with no threats to their hegemony. In order to do this, they needed to maintain the principles of hierarchy and social control established in the colonial era (Butler 1998).

At the same time, a solid body of historical research on race relations shows that even under the dehumanizing conditions of slavery, some blacks owned property, were literate, and had considerable control over their work (Borges 1992; De Queirós Mattoso 1992; Bacelar 2001). Therefore, even within a highly unfavourable context, some Afro-Brazilians managed to improve their lives. Thus, the question of the integration of the black individual in Brazilian society has long been a concern in debates about social relations in Brazil (Cardoso and Ianni 1960; Fernandes 1969; Guimarães 1987; Ianni 1966). Raymundo Nina Rodrigues, trained in Bahia as a physician, published the first major analysis of Afro-Brazilians and their role in national society in the late nineteenth century.[10] Rodrigues' assessment (echoed by other scholars of his time) was that African retentions had had a profoundly negative effect on the ability of black people to assimilate into mainstream Brazilian culture. Influenced by the perspectives of criminological psychiatry, he concluded that Afro-Brazilian culture was a manifestation of racial evolutionary backwardness.

The publishing of Gilberto Freyre's *Casa Grande e Senzala* (*The Masters and the Slaves*) in 1933 shifted this narrative by arguing that the ideas and relationships shaped during slavery (not African retentions) played a decisive role in determining race relations in Brazil. Freyre presented a social history of the Brazilian family that was set against the backdrop of a patriarchal rural and feudal system organized around the sugar plantation economy and slave labour. He gave colonial Brazilian race relations a more

[9] See the work of Sônia Santos, "*Brazilian Black Women's NGOs and their Struggles in the Areas of Sexual and Reproductive Health*"; Jaime Alvez, "*Narratives of Violence: the White ImagiNation and the Making of Black Masculinity in City of God*"; and Luciane Rocha, "*Research(ing/in) State Genocide: Toward an Activist and Black Diasporic Feminist Approach.*"

[10] Rodrigues published the first instalment of *O Animismo Fetichista dos Negros Bahianos* in the *Revista Brasileira* between April and September 1896, followed by a final instalment in volume 9 of the same journal in 1897. These articles were subsequently published in *L'Animisme Fétichiste des Nègres de Bahia* in 1900. Raymundo Nina Rodrigues, *O Animismo Fetichista dos Negros Bahianos* (Rio de Janeiro: Civilização Brasileira, 1935), 7–8.

positive spin by focusing on interracial and cultural mixture between white, indigenous, and black people, which became central to the concept of Brazil as a racial democracy. Both works paved the way for further sociological and anthropological research on race relations in Brazil, in which the Northeast—especially Bahia—featured as a privileged location that exemplified the ideals of harmonious racial relations. Thus, classic studies on race relations in Bahia generally conformed to the dominant narrative of the state's supposedly "cordial" race relations (Pierson 1942; Herskovits 1943; Frazier 1942; Landes 1947). Indeed, the idea that an individual of African descent could succeed in Brazil because of the country's lack of prejudice was often used as a key plank in arguments supporting racial democracy, even as most blacks remained poor.[11]

Nevertheless, the supposed inability of black Brazilians to assimilate into a modernizing and industrializing Brazil remained a persistent concern. Brazilian sociologist Florestan Fernandes produced a classic work on race relations in Brazil—*A Integração do Negro na Sociedade de Classes* (1965; translated as *The Negro in Brazilian Society*, 1969)—based on research he carried out in São Paulo. In this book, Fernandes set out to explain the relative inability of Afro-Brazilians to attain upward mobility in a modern economy. He concluded that the social problems plaguing blacks in the late twentieth century were the result of the legacy of slavery and their inadequate adaptation to an industrializing, capitalist society.[12] Thus, he gave little weight to contemporary racial prejudice and discrimination.

Thales de Azevedo entered this circle of debate in the early 1950s, beginning an intense dialogue with scholars such as Donald Pierson, Charles Wagley, Costa Pinto, Guerreiro Ramos, Marvin Harris, Roger Bastide, Florestan Fernandes, and Octávio Ianni, among others. De Azevedo's ethnographic research on *elites de cor* (colour elites) in Salvador, first published in French in 1953 as part of a UNESCO project on Brazilian race relations, focused on the intersection of class, status, and colour—particularly the role of *mestiçagem* (racial and cultural mixture) in enabling

[11] This idea was echoed by the more widely known Joaquim Nabuco, "Color in Brazil is not, as in the United States, a social prejudice against whose persistence no character, talent or merit can prevail" (quoted in Skidmore 1974, 24).

[12] Fernandes' thesis has been challenged by numerous scholars. For example, the Brazilian sociologist Carlos Hasenbalg, writing not long after Fernandes, charged that slavery alone could not explain the subordinate position of blacks in Brazilian society and that racial discrimination was also to blame (see Butler 1998).

social mobility for *elites de cor* in Bahia. De Azevedo emphasized a tacit agreement between blacks, whites, and coloured Bahians to keep racial prejudice at a minimum (De Azevedo 1996, 49). He believed that strategies of "whitening" among *elites de cor* led to some degree of integration. As whiteness is associated with wealth, higher social status, and prestige, upwardly mobile blacks could symbolically "whiten" through acquiring money and education or through marrying light-skinned spouses, producing even lighter-skinned children. His ethnography demonstrates the different pathways of social ascension available to non-white Bahians in spheres such as marriage, commerce, politics, civil service, the military, the arts, education, religion, sports, and the liberal professions.

However, Thales de Azevedo acknowledged the limits of *mestiçagem* as a mechanism of upward mobility, such as the persistence of the importance of social origin, family, and colour in Brazilian social relations and the relatively rigid tendency to marry within one's social group. He also noted that aligning oneself with whiteness did not necessarily protect from racism. For example, he pointed to the widespread belief among professionals of colour (such as teachers, doctors, church leaders, and public servants) that in order to be successful, one had to work twice as hard and be "double the value of your white competitor" (De Azevedo 1996, 110).

Race, Social Mobility, and the Brazilian Family

Private family dynamics help to construct, reproduce, and reinforce an unequal Brazilian society. Brazil has long been considered a patriarchal society in which men control households, businesses, and politics and women committed to fidelity and domestic labour (Medeiros 2018, 25). More recent work on Brazilian gender relations, however, has sought to "recover" the agency of women, who despite being more or less confined to the domestic realm may assert themselves "through sexual or emotional influence, in order to participate in the decisions, acquire material possessions, circumvent (male) authority; and get things done in their own way" (Corrêa 1993; Parker 1991; Rebhun 1999; Machado Borges 2003, 200).

Among the most popular representations of Brazilian family life—which linked the home to broader social relations—are those constructed by Gilberto Freyre, in which the social structure of the plantation house in *Casa Grande e Senzala* (1986) took centre stage. The violent reality of slavery was replaced with a fairy tale of racial harmony and unity, rooted in the idea of Luso-tropicalism in which Portuguese colonialism and slavery were framed as somehow kinder than in other European colonies, leading to interracial sexuality and intimacy in the Brazilian family. This process of racial and cultural mixing, or *mestiçagem*, was seen to be the foundation of the Brazilian nation. For Freyre, this process unfolded in the extended families of the Euro-Brazilian elite, which stretched out to include illegitimate members of the family and dependents of all sorts, enabling the incorporation of members across lines of race and class:

> Every Brazilian, even the light skinned, fair haired one, carries about with him in his soul, when not on soul and body alike—for there are many in Brazil with the mongrel mark of the genipap—the shadow, or at least the birthmark, of the aborigine or the Negro. Along the seaboard, from Maranhão to Rio Grande do Sul, it is chiefly the Negro. The influence of the African, either direct or vague and remote. (Freyre [1933] 1986, 278)

While Freyre's writings have since been critiqued, historical studies suggest that a high degree of consensual unions has long been the norm in Brazil (De Queirós Mattoso 1992; Borges 1992). This representation of Brazilian social life fails to recognize the unequal relations that unfolded within the intimate spheres of the family. Indeed, for many families, axioms of servitude, patronage, and dependence have long anchored social relations, while masking power relations. For example, the bond of godparenthood and informal ties of kinship may have offered protection to some but excluded others (the term *protegido*, or "protected one," refers to the dependents of patrons, suggesting that it was risky to navigate society without one). Parents of varied social backgrounds often sought out individuals who could benefit their child in some way. According to the historian Katia de Queirós Mattoso, "In a society in which positions depended on the help of third parties, the choice of a well-connected godfather or godmother was part of a strategy of social ascent, or at least, the preservation of a condition already attained" (1992, 175). Within such a system, the support of a patron would have been essential for a free person of colour, who could hardly expect to aspire to high positions in the civil service or military without recommendations and references from well-positioned patrons.

Patronage as a mode of social interaction and possible strategy of social ascension was a dimension of Brazilian culture which was accepted by (or forced upon) Afro-Brazilians, not only conditioning their relationships with whites and the larger society, but also informing the structure of relationships within the black milieu. For example, some Afro-Brazilians were important patrons (though limited in scope) within their communities, through long-standing traditions of black philanthropy practices through organizations such as Afro-Brazilian brotherhoods and other social institutions such as trade organizations or benefit associations (Butler 1998, 22–23).

Practices of manumission in Brazil offer convincing evidence of preferential treatment within the extended family, where mixed race slaves (*pardos/mulatos*), especially children, enjoyed advantages over African-born and unmixed slaves. The American sociologist E. Franklin Frazier—a leading scholar of race relations and the black family—wrote that:

> Under the patriarchal organization, the Portuguese and the Negro slaves lived in close and intimate association, a large mass of mixed bloods came into existence who enjoyed special privileges because of their kinship with the master class. (1942, 466)

Mixed raced children were often the offspring of their mother's master and so affective bonds may have afforded them more privileges.[13] For example, mixed race slaves often worked in the *Casa Grande* (Big House) and were seen by others as privileged for doing so, as they avoided the back-breaking work of the field slaves. They were more likely to incorporate "white" elite mannerisms, language, and customs. Some *pardos* benefited from a good inheritance or the good will of their white fathers and some received an education.

However, a focus on the privileges of mixed-race children can sometimes obscure the fact that the family mirrored the hierarchy of the dominant society. The social norms of a brutal slave society were forced unto enslaved and free African descendants alike, influencing patterns of thought and action which have endured or shifted over time, revealing both accommodation and resistance. For example, *pardos* and *mulatos* were involved in both active and passive resistance in the Master's house,[14] joined with free blacks to form religious brotherhoods and mutual aid organizations, helped enslaved peoples escape the plantation, and supported revolts of enslaved Africans.[15] Nevertheless, *pardo* and *mulato* chil-

[13] See Meiko Nishida, "Manumission and Ethnicity in Urban Slavery," 386.

[14] See Sandra Lauderdale Graham, "House and Street: The domestic World of Servants and Masters in Nineteenth Century Rio de Janeiro."

[15] See Stuart B. Schwartz, "Manumission of Slaves," 635–636.

dren were often framed as superior to their mother's unmixed children.[16] In a context like this, enslaved women often relied on their sexuality to attain social mobility and freedom for their children (Hordge-Freeman 2015, 12).

Another key strategy was to acquire a family name. In order to give an illegitimate child a better life, he or she might be given the surname of the father, especially if he was well known and had some social status or was lighter-skinned or white. In other cases, parents tried to cover up their children's mixed-race ancestry by stating in baptismal records that their children were white (Reis and Sabrina Hill 2003, 30). White fathers sometimes recognized their offspring born of affairs with black and mixed-race women and in some cases, they were legally recognized and inherited part or all the assets of their father. The Brazilian writer, Machado de Assis, wrote about complex racial situations like this in the Bahian family in his novel *Helena*, first published in 1876. The heroine of this novel is a brown-skinned young woman of obscure birth who has two possible fathers. She is introduced in the book as the illegitimate daughter of the late Conselheiro Vale, from an upstanding, educated, white family. An old family friend of Vale informs the Conselheiro's adult children, with great difficulty, of the embarrassing news that the father had been indiscreet and fathered a child with a black woman. This child, who appears in the novel as Helena, was educated at the Colégio de Botafogo and was to inherit a part of Conselheiro's wealth. It was also his wish that the child come to live with his family, as if she were of their blood. This news was shocking to the Conselheiro's family. The Conselheiro's legitimate daughter by marriage preferred to place Helena's inheritance at her door, being against the intrusion of an unexpected addition to the family from someone who she felt had no rights and should not be legitimized in the eyes of others as a family member.

However, as the story goes, the son feels obliged to bring Helena home, despite the sister's vehement opposition. When Helena comes to live with them, they are gradually drawn to her warm personality. Helena is mysterious—outwardly, she developed ways to cope by cultivating an

[16] In a key debate that has influenced race studies in Brazil, Carl Degler suggested that the offspring of black and white Brazilians had access to privileges through a "mulatto escape hatch." A decade later, Carlos Hasenbalg and Nelson do Valle Silva suggested that outcomes over the lifetime for browns and blacks are relatively similar. See Degler, *Neither Black nor White*; Hasenbalg and Silva, "Educação e diferenças raciais na mobilidade ocupacional no Brasil."

ability for dissimulation that allowed her to overcome both her present social position and the inherent sense of shame linked to the past. However, as the story unfolds, we see Helena's interest in a particular house and its strange occupant, who eventually is revealed as her biological father. The story therefore suggests that Helena's mother had named Vale as the biological father so that her daughter would have a chance of a better life. The point being made here is that social mobility for African descendants often depended on making a relationship with the Euro-Brazilian elite, whether through blood ties, marriage, or fictive kinship.

Transformations After Slavery

Historical studies show that the racial scenario after the abolition of slavery did not change much for African descendants. There was a strong anti-Africanism sentiment among the conservative Euro-Bahian elite, concretized into legislation that made it difficult even for free persons of colour to improve their lives (Butler 1998; Guimarães 1982). A large part of the economically active population comprised African descendants, most of whom had little education and worked in areas such as personal services, crafts and skilled work, civil construction, and small commerce (De Almeida 2006). Black women continued to be associated with domestic work in the homes of local elites, working as nannies, cooks, and washerwomen. However, a special group of workers—called *operários* and *artistas* (factory workers and craftsmen)—began to take shape in the late nineteenth and early twentieth centuries (Guimarães 1987; Guimarães et al. 1995). Figure 2.5 is a portrait of a group of porters and dockworkers likely taken in the Lower City, Salvador, at the turn of the twentieth century.

They included textile mill workers, stone masons, stevedores, cart, trolley and taxi drivers, typesetters, waiters in restaurants, and employees of small workshops (Borges 1992). Sanctioned means of ascent, though limited, did exist for black and mixed-race people, so that some gained considerable occupational mobility, occasionally as artists, politicians, and writers. Aside from income and property, claims to "dignity" and outward signs and symbols such as dress, education, and marriage to lighter-skinned persons distinguished the "middle sectors" from the workers (Borges 1992, 34).

Despite important exceptions, Brazil's hierarchical system made it difficult for Afro-Bahians to turn opportunity into long-term, durable

Fig. 2.5 Group portrait of porters, Bahia, Brazil. (Source: From the New York Public Library, https://digitalcollections.nypl.org/items/b91b64cc-f5dc-da78-e040-e00a180657a2)

upward mobility. For example, Mahoney's case study of Afro-Brazilians and social mobility in the cacao growing area of southern Bahia in the late nineteenth to early twentieth centuries suggests that while some were able to acquire land and begin to exploit it, commercial, legal, and political structures made it difficult for them to maintain control of their properties and subsequently pass on their status to their children (Mahony 1997).[17]

Whether one resided in the rural or urban areas, Brazilians navigated an unequal society as they lived through the fall of the empire and the establishment of the first Brazilian republic. The economy of the Northeast—despite some diversification—began to decline as it faced competition

[17] Brazilian inheritance laws tended to break up property in successive generations, working against the ability of many small farmers (Afro-Brazilians included) to hold onto their property. Other factors, such as the racial tensions and the social distance between Afro-Brazilian small farmers and the merchants, farmers, planters, politicians, and bureaucrats with whom they dealt could have also affected their ability to take advantage of the opportunities of the boom in cacao production in southern Bahia during this time.

from the sugar-producing Caribbean and the coffee-producing areas of southern Brazil (Mayblin 2010; Muniz de Albuquerque 1999). Unequal land distribution further contributed to regional inequality and poverty in the Northeast (Mayblin 2010; Pereira 1997). These historical-political and economic events contributed to social and structural inequality, which in turn had effects on the lives of Afro-Brazilians.

Recent Transformations

The 1950s represented the beginning of a radical transformation in the socioeconomic life of Salvador with the discovery of oil in the Bahian hinterlands. A heightened phase of urbanization and industrialization began, leading to the establishment of modern companies in the Aratú Industrial Centre and later in the Camçari Petrochemical Complex (De Almeida 2006; Guimaraes et al. 1995; Agier 1994). The impact on the economy of Bahia was intense, with repercussions on the lives of African descendants (Bacelar 1999, 95). Petroleum exploration led to the establishment of Petrobras (the state-owned Brazilian petroleum company), transforming the Recôncavo and contributing to the modernization of the city, including in its commercial and financial sectors (De Almeida 2006). As the city experienced an economic boom from oil extraction, new jobs and demand combined to bring about an insertion of many Afro-Bahians into this sphere (Guimarães et al. 1995). This was crucial to the creation of a stable black working class in the city.

Education became a sanctioned means of social ascent for many African descendants throughout the twentieth century, especially because of the emphasis on educational reforms and expansion under Vargas's Estado Novo (1937–1945). By the 1950s, De Azevedo noted that there were many teachers of colour who commanded an enormous amount of respect amongst their students. Education became a powerful symbol of distinction and a way to distance oneself from the poorer masses and from blackness. African descendants made great efforts to send their children to elementary education and went through even greater sacrifices to maintain them in secondary school education, to give them a chance at a better life (De Azevedo 1996, 109).

Meanwhile, Salvador underwent marked modernization. With the appointment of the politician Antonio Carlos Magalhães (henceforth, ACM) as Prefect in 1967, a large amount of federal money became available for public works in the city (Neto 2006; Souza 2000). There was a massive sale of public land which was snapped up by developers (Neto 2006, 332–338). In 1971, the Central Business District (CBD) was created, where the government implanted its administrative nucleus along the principal Avenida Paralela (Neto 2006, 400–431). The creation of the CBD stimulated the development of the new business district, linking it to the transportation network and the airport. Several new buildings were constructed in this area, such as Shopping Iguatemi, a luxurious mall, constructed in 1975; a hypermarket; and a new central bus station built around the same period. Along the Orla, *bairros nobres* (new, modern neighbourhoods) flourished with the construction of ultra-modern condominiums for Salvador's up and coming business and professional classes. Services sprung up to serve these neighbourhoods and there has been a considerable investment in infrastructure, telecommunications, and recreational activities. New industrial infrastructural projects, such as hydroelectric plants, were built outside of the city, bringing needed economic development to Salvador and Bahia more generally, linking the state to international tourism and consumer networks (Alberto 2011, 283).

Despite the diversification of the economy and modernization of the city, Brazil was still a country of deep contradictions. Racial and gender hierarchy continued, as well as old beliefs that connected blackness to inferiority and whiteness to superiority. Brazil entered a period of military dictatorship from 1964–1985 during which the idea that Brazil was a racially tolerant nation gained a strong foothold. Propaganda by mostly white men upholding the dictatorship involved the dissemination of a unified image of Brazil. During this period, racism was not to be discussed and critiques of racial democracy (considered as "acts of subversion") were not to be tolerated (Nobles 2000, 111). Despite the suppression by the government of racial discourse, by the 1970s a new era of black mobilization emerged.

In Bahia, working and middle-class blacks were attracted to black solidarity messages from within and outside of Brazil. Inspired by the North American Civil Rights and Black Consciousness movement, many sported Afros, platform shoes, and Black-Power fashion and listened to soul music from the United States (Dixon 2016, 44). Bahia developed its own local soul and funk music known as "Black Bahia," giving rise to musical forms associated with racial consciousness, including carnival groups engaged in a new kind of radical black politics (Alberto 2011, 285). This was followed by the re-Africanization of carnival in 1975, setting the stage for a new era of racial consciousness and social activism.

The 1980s were marked politically by Brazil's democratic opening, increased liberalism, and more "enlightened" policies regarding race (Dávila 2012). At the same time, black feminists struggled to create a new set of symbolic associations for black women beyond the confines of stereotypical representations, such as the black domestic worker. However, a severe economic crisis began in 1985, extending into the presidency of Collor de Mello, who was the first to be democratically elected after the end of the Brazilian military government. The economic crisis lasted into the 1990s, characterized by reduced foreign investment, extreme inflation, and high unemployment (O'Dougherty 2002). Following a neoliberal economic plan, the *Plano Collor* was introduced in order to stabilize the economy, but this led to a worsening of the well-being of the working classes and increased inequality. Subsequently, President Fernando Henrique Cardoso (1994–2003)—a sociologist whose main research dealt with the socioeconomic marginalization of Afro-Brazilians—expanded the space for social movements advocating for affirmative action to thrive. His economic reforms included the implementation of the *Plano Real*, introducing a new currency (the *real*), which temporarily benefitted some before the economy destabilized again, although less severely, from 2001–2003 (Mendes 2015).

The 1990s and early 2000s represented a watershed moment in the country for tackling racial inequality through a vigorous debate on affirmative action policies. Afro-Brazilian civil society groups and politicians had long called for measures to address the lack of blacks in the Brazilian university system, the poor quality of public education, and the low position of blacks in the labour market. These debates, discussions, and activism by the Brazilian Black Movement led to important policy changes resulting in the implementation of affirmative action programmes across Brazil, challenging social and political relations that had long been deeply racialized and hierarchical (Martins et al. 2004).

President Fernando Henriques made a key move in opening the debate surrounding affirmative action in 1996, leading to a range of uncomfortable questions and reflections on the intersection of race, social justice, education, and citizenship with respect to black, indigenous, and poor Brazilians (Telles 2004). These reflections represented a critique of established Brazilian social relations, suggesting that blacks had been the historical victims of deeply embedded structural racism. Subsequently, key events included the adoption of affirmative action by the Federal University of Rio de Janeiro in 2000 and 2001 (Cicalo 2012; Lima 2012); the enactment of a federal scholarship programme in 2005 aimed at poor, black, and indigenous students (referred to as the University for All Program, or Pro-Uni; Lima 2012); two ground-breaking legal victories by the Brazilian Supreme Court in April and May 2012 respectively; and the signing of a comprehensive law known as the law of Social Quotas by President Rousseff (Dixon 2016).

In 2002, Luiz Inácio Lula da Silva of the Workers' Party (Partido dos Trabalhadores) became Brazil's first socialist president. Building on previous social programmes designed to reduce poverty and long-term structural inequality, the two administrations of President Lula brought improved management and coordination to these government schemes, implementing the Programa Fome Zero (Zero Hunger Program) and the Bolsa Família (Family Stipend). The latter programme is one of the largest cash transfer programmes in the world, benefitting around 50 million people or a quarter of Brazil's population (Vargas 2016, 553). It particularly benefitted the Northeast and is credited with lifting many non-white Bahians out of poverty (Ansell 2014). Nationally, the Bolsa Familia Program enabled 31 per cent of families that had been living in a condition of extreme poverty to escape that condition in 2006, although only 4.9 per cent rose above the poverty line as a result of the impact of this programme (Ivo 2008, 194). Critics of the Lula administration argue that anti-poverty programmes based on conditional cash transfers have little long-term structural impact on social inequality. However, the scope of these programmes was greater under the Lula administration than previous administrations and their impact on poverty has proved sustainable.

Other social programmes also had an impact on inequality, such as targeted pension benefits and unemployment insurance.[18] Data on income revealed significant transformations during this period of pro-poor political commitment—between 2001 and 2009, income increased to 69.08 per cent among the 10 per cent poorer and increased to 12.80 per cent for the 10 per cent richer (Vargas 2016, 554). Furthermore, the Brazilian economist Marcelo Neri stated that in 2001, black incomes were 53 per cent of white incomes. By 2009, black incomes jumped to 62 per cent of white incomes (Neri 2011, 15). Meanwhile, the historical subsidies that the middle and upper classes systematically received from the government

[18] The Gini Coefficient of inequality in income distribution in Brazil was 0.604 in the early years of the 1990s, falling to 0.587 by the end of Cardoso's administration and reaching 0.5687 by the end of 2004 (Ivo 2008, 191).

in tax breaks and education contracted, which contributed to the Workers' Party losing some of their historical middle-class, student, and organized labour bases (Vargas 2016, 553). Indeed, in more recent years the middle classes have tended to shift away from Lula and his party (Vargas 2016, 554).

Despite an expanding support base in the country's non-white, urban, mostly informal labour force, Lula's party revealed an ambiguous relationship to blacks. For example, blacks remained underrepresented in politics (Dixon 2016) while others were frustrated with changes that seemed to be largely symbolic. While Lula created the Secretariat of Policies for the Promotion of Racial Equality (SEPPIR) in 2003 and gave it ministry status, Luiza Bairros—a seasoned Afro-Brazilian activist who headed the ministry after Matilde Ribeiro resigned in 2016—complained of the ministry's toothlessness and bureaucratic red tape (Vargas 2016, 556).

A period of strong economic growth between 2004 and 2009 propelled 4.8 million Brazilians out of poverty, reducing the proportion of poor people in six principal metropolitan regions (Gledhill and Hita 2009, 4). However, this was followed by economic recession and political instability, including the 2013 massive street protests across several Brazilian cities, the impeachment of Dilma Rousseff in 2016, and the rise of the Far Right in Brazil. This book will not focus much on this later period, as the majority of my fieldwork was carried out between 2005 and 2007, with short stays in the country up until 2017.

Over the past three decades, modernization processes in Brazil—such as rapid industrialization, increased rural to urban migration, and the growth of the urban middle classes—have resulted in the reorganization of class structures and gender roles, including an increase in women's employment in the formal sector (Besse 1996; Sanabria 2016).[19] Black women, however, have long worked outside the home in informal sector work. Meanwhile, Salvador's population has almost doubled in size.[20] This population then became increasingly absorbed into the tertiary sector through the performance of low-paid, non-contractual, and casual services (De Souza 1979: 47–48). As women began to work in factories and service positions, the feminization of the labour force contributed to the

[19] From 1970 to 1990, women's economic participation rate in the formal labour force has more than doubled (Edmonds 2010).

[20] The years 1940–1990 witnessed a dramatic increase in population from 400,000 to more than 200,000 inhabitants (Souza, 1979).

redefinition of gender roles and household power dynamics (Medeiros 2018).

The petrochemical complex in Bahia, which began operations in 1978, did not have the long-term impact that urban planners expected it would have on Salvador's economic development. For example, for every three directly generated jobs in 1990, it cost a million dollars in investment—and the investments were mostly from outside Bahia, which is also where most of the profits went (De Almeida 2006, 25). As a result, it has been the tertiary sectors—such as tourism, financial and commercial services, and public administration—which have been the leading employment sectors in Salvador's pattern of modernization.

In sum, it is important to recognize the ways in which these larger politico-economic and social transformations intersect with historically entrenched structural relations, impacting the trajectories, experiences, and identities of Afro-Brazilians across the different generations in the families I got to know in Salvador. In the following sections, I aim to bring these people, their experiences, and their voices more into focus.

Traversing Neighbourhoods in Salvador

Neighbourhood spaces have long been popular sites for anthropological research, as they are seen to play a role in structuring urban belonging and everyday experiences of the city (Jaffe 2016). Traditionally, urban anthropological research—such as the classic case studies of the Chicago School—focused on marginalized groups and neighbourhoods as well as ethnic enclaves.[21] However, my research participants came from different neighbourhoods and were not necessarily concentrated in black neighbourhoods. Some research participants lived, or had lived in the past, in low- and middle-income neighbourhoods primarily associated with having a large black population, such as Liberdade and Garcia. Others lived in high-rise apartment buildings located in higher income neighbourhoods such as Graça, Ondina, or Barra. Figure 2.6 shows a quiet neighbourhood near to a university campus in Federação. Figure 2.7 is a photograph taken by the author from an apartment window in Barra showing some of the neighbourhood.

[21] In the United States, work on class included the community studies by scholars such as Warner (1959) Warner and Lunt (1941, 1942), Goldschmidt (1950), and others.

Fig. 2.6 Apartment buildings on a quiet street in Federação, Salvador. (Source: Doreen Gordon)

Liberdade is an old, working-class neighbourhood that is famous for its strong sense of black culture and identity and is well known throughout the city for its cultural activities, such as its funk dance parties, carnival activities and for Ilê Aiyê—a *bloco afro* (Afro-block) carnival group.[22] Liberdade has been able to capitalize to some extent from this reputation. It was different to the other neighbourhoods I visited and lived in, as a great deal of socializing took place in the community where some families had lived for generations and had life-long friends. Figure 2.8 shows the Curuzu district in the Liberdade neighbourhood where Ilê Aiyê, a famous *bloco afro* (carnival group), is located.

I was close with a family headed by a single mother in her 50s, Norma Sanchez, and her two adult children, Bia and Danilo, who had gradually worked their way through the educational system at great sacrifice and settled into stable jobs. They experienced considerable mobility, despite various obstacles along the way, and have since left Liberdade. Today,

[22] Referred to as a *bloco afro*, it features themes from global black culture and history and celebrates the aesthetic beauty of black people.

Fig. 2.7 View from an apartment in Barra neighbourhood. (Source: Doreen Gordon)

Danilo is married and has a job in the police force while he is studying for a law degree in the evenings. Meanwhile, Bia has a senior management position in a bank and since 2012, has been living in a city on the outskirts of Salvador called Laura de Freitas, in a townhouse in a self-sufficient, gated community. While the area has a lot to offer in terms of restaurants, malls, and recreational activities, it is a long commute to get into the city centre for work.

Other research participants lived, or aspired to live, in places like Campo Grande, Brotas, Federação, Canela, Barra, Rio Vermelho, Ondina, Pítuba, or Itapuã—and in gated communities far from the city centre in coastal areas like Stella Maris. These neighbourhoods are associated with having well-organized streets and services and higher social status. Traditionally, middle-class houses had a *quintal*—a backyard with several fruit trees, vegetables, and herbs. The Teixeira family—members of whom I met early on in my fieldwork through a small circle of friends and acquaintances—had a property like this which had been inherited through the generations on the mother's side of the family. However, nowadays, it is more

Fig. 2.8 Curuzu in the Liberdade neighbourhood

common for middle-class people to live in apartment buildings—though there were a range of different kinds. There were the older, more traditional ones; the more modern but simply built buildings and the newer, super upmarket ones, replete with luxury materials such as granite and marble, pretty verandas, stylish garden patches, and in the shared public areas, swimming pools, party halls, sports areas, and a tiny concrete shelter for a porter. These self-contained, luxury apartment buildings typically have several stories of garages. The apartments themselves are small, but in the eyes of residents, the lack of space is compensated for by the agreeable sensation that those living there belong to a new world: one that is modern, hygienic, privileged, stable, and secure. Figure 2.9 gives an idea of what some of the newer luxury apartments look like in Salvador.

Apartment buildings are often managed by a *síndico*, or building manager, who is an important feature of middle-/upper-class life in this context—he or she regulates and organizes things, sets the rules and norms of behaviour, and gives final approval for parties/social activities which are held in the common lobby area. The building manager often sets the tone for the entire building. One of my research participants lived in an

Fig. 2.9 Luxury apartments in Salvador

apartment building where the *síndico* planned a gay wedding in the lobby area—this apartment building was known as "alternative," which is why she liked it in comparison to the more mainstream, white middle-class apartment buildings where the *síndicos* and residents are more conservative.

The community or neighbourhood became a less important and relevant location for socialization in higher income neighbourhoods. Residents were more likely to create their own meaningful spaces across the city, socializing with friends and family in their own apartments, going out to restaurants, bars, and places of cultural interest (museums, art galleries, theatres), watching a movie at the upscale shopping malls, spending time on the university campuses, or attending private parties. Some people hardly came out of their neighbourhood—for example, one research participant lived in Pituba and said that everything her family needed was there. However, she made special efforts to take her children into areas of the city they did not know as well, such as Ribeira.

Brotas is another highly populated area of the city in which quite a few research participants lived or spent some time living. It is densely populated by apartment buildings, houses, businesses, schools, hospitals, and popular neighbourhoods. Several middle-class apartment complexes were constructed here as well as more expensive, high-security, luxurious looking apartment buildings. In the past, this neighbourhood, as well as Itapagipe where the famous Bom Fim Church (The Church of Good End) is located, had similar socioeconomic profiles, were important agricultural and fishing areas, and played important cultural roles in the city. Their large black populations and location on the edge of the city favoured the development of *quilombos* (maroon communities) and Candomblé temples (Reis and Sabrina Hill 2003, 20). Two respondents whom I interviewed lived here. One was DJ Funk Soul, who was a bachelor living in his own apartment. While he enjoyed some aspects of living in Salvador, he noted that:

> Salvador is the black capital of Brazil, but they do not valorize the black person here, or even black music. There isn't that much diversity in black music, or in radio or television programmes. They talk about negritude here, but it does not exist. Maybe only in places such as Pelourinho (the historical center). Instead, when I walk on the street with my big afro, people are staring at me, like I am an animal. I do not know if they do not like it, or they just think it's strange. If you have big hair… colourful clothes… people think it's strange. People here are very prejudiced. You can't dress as you like here. You can't be different because you will call attention to yourself. Salvador is like a grand rural town—it is very conservative. (Interview: May 9, 2006)

Another research participant, Horacio Rocha, who worked as a highly skilled transport engineer at the train station in the Lower City, was divorced and lived alone in Brotas in a modern apartment building. He had two adult children just starting university who also lived in Brotas in their own apartment. Horacio has fond memories of the neighbourhood where he grew up. His parents were solid working-class people—his father a lower level technician working in civil aviation and his mother an administrative employee of a public school. They lived on a street in Garcia with other working-class families whose members had jobs on the dock or in factories. He stated that this neighbourhood had a strong community spirit and a sense of cultural traditions. In addition, the families on his street knew each other well and were active in their communities. The neighbourhood had a square, with theatres and a cinema, regular festive gatherings, carnival parades, churches, and community football events, thus providing opportunities to forge close neighbourhood bonds. He missed this aspect of community life in Brotas, as life is more impersonal and private in the high-income neighbourhood in which he now lives. He maintains involvement in his former neighbourhood through his membership in a Baptist church in the area which provides him with a base of informal friendships. He described the church as very plural and as having a strong presence of *negros da classe alta* (well-off black people) as well as *gente de baixa renda* (lower income people) and *operários* (working-class people). This suggests that some types of associational life help to mediate class differences in Salvador. Figure 2.10 illustrates a mixed neighbourhood in Federação, with apartment and houses on the higher level and popular neighbourhoods in the valley.

Everyday Racism in the "Black Rome" of Brazil

How did Soterpolitanos—as the residents of Salvador are sometimes called—view racism? Generally, there is greater awareness of the different forms of racism in Brazil and increased mobilization around issues of race in recent years. Most research participants considered everyday interracial situations in the public sphere as well as within the intimate sphere of the family, revealing a repertoire of negotiation strategies including humour, discursive practices, direct confrontation, denial, and ideological innovations. For example, a common joke often related to me was about two boys who were friends—one black and one white—who were running to catch a bus in the busy historical centre of Pelourinho. A policeman, who was standing on the street corner, watched the white boy run past him but when he saw the black boy behind him, he pounced on him, assuming that he was trying to rob the other one. The white boy stopped to explain that they were running to catch the bus, but the policeman did not believe him. On another day, this same policeman witnesses another black boy

Fig. 2.10 Mixed neighbourhood in Salvador—apartment buildings surrounded by popular neighbourhoods. (Source: Doreen Gordon)

running past him, followed by a white boy, and exclaims with a smile, *os meninos estão brincando muito!* (the boys are playing a lot!).

Other research participants described stereotypical media representations of blacks in films and *novelas* (soap operas), or racism within the workplace. For example, one research participant, Jacinda Figueiredo, described her working life in Brazil as one of consistently moving into better positions, but encountering racism when it came to senior positions. The eldest daughter of her parents, Jacinda had completed a course in tourism in high school. She began her working life as an office cleaner at the Hotel da Bahia. The general manager of the Hotel da Bahia eventually noticed her work ethic and she was promoted to assistant secretary, Human Relations Department—a job that opened the way for her to participate in training workshops and to learn other skills. She received praises from senior management for her steady development, organizational skills, and confidence in dealing with a wide variety of people. Her upward career trajectory at the hotel included being promoted to secretary, General Finances Department, and personal assistant to the general manager. However, she noticed that after nine years, the general manager had no intention of promoting her to any other position within the hotel, so she decided to leave.

Reflecting on this situation, she explained that most of the staff of the distinguished hotel were light-skinned, especially those interacting intensely with customers, in order to portray an image of a "modern" city to tourists. Nevertheless, her former managers gave her glowing recommendations when she applied for a job at Varig airlines as an air hostess. Jacinda admitted to being somewhat nervous at the interviews when she noticed that she was the only black candidate. However, her work experience impressed her interviewers and she was offered a job as a customer relations agent, attending to clients in the airport in Salvador. She surmised that she had been assigned this job because the glamourous position of air hostess required university-educated females with a certain appearance (light-skinned, long hair, European-like features). Nevertheless, she took the job and excelled at it. After a few months, a friend encouraged her to apply for a similar position with the same airline in Porto Alegre. She got the job and moved to Porto Alegre, where she lived as a single woman in a comfortable apartment and went on vacations to Latin America and Europe. However, one day she was asked to temporarily fill the position of her general manager who was ill. She noticed that the workers were uncomfortable with this. They began spreading rumours and gossip about her—such as her closeness to senior management staff and her lack of awareness of her "proper place" in the world of whites, especially as a black woman from the Northeast—but Jacinda said that she had always had a strong work ethic and a confident personality, with the ability to interact with people of different backgrounds. Therefore, she simply paid attention to her job, where she continued to get positive reviews. She was soon asked to be the head of the Lost and Found Department at the airport. Eventually, a German friend whom she had helped in the airport invited Jacinda to Germany on vacation, and she met her future husband there. He encouraged her to apply for a job with Lufthansa airlines. She got the position and now lives in Germany with her husband, visiting Brazil occasionally to see her family. While racism did not prevent Jacinda from achieving considerable career mobility, she encountered racism and sexism at the higher occupational levels despite her obvious abilities and talents.

Laws against racism in Brazil have existed for some time, such as the Afonso Arinos Law of 1951, passed after an African American dancer, Katherine Dunham, was denied a room at an upmarket São Paulo hotel. Furthermore, Brazil's 1988 constitution defines racial discrimination as a crime without bail or statute of limitations and punishable by imprisonment (Dávila 2012). During my fieldwork in Brazil, I personally knew people who had filed cases of racial discrimination against others or who were aware that they could do so. In one case, Rosa Figueiredo—Jacinda's mother—related a story about how she had sent Jonas, her second youngest child, to the grocery store to buy a few things for the house. He returned home minutes later, crying. She asked him what had happened. He eventually told her that the cashier had embarrassed him in front of all the customers, looking at his clothes and thinking that he could not buy the groceries. Angrily taking hold of her son's hand, Rosa marched up the steep hill from their house to the small neighbourhood grocery store. Jonas led her to the cashier and she spoke calmly but firmly to the woman, saying that she had sent her son with sufficient money to buy some groceries and that she did not have the right to speak to him in an insulting manner, especially as he was only eight years old. Rosa explained that she was teaching him to believe that their society was scornful of black people, yet he was a perfectly respectable boy. She went on to accuse the woman of racism and asked to speak with the manager of the store. She explained the situation to the manager, while at the same time attracting quite a crowd of customers who had gathered to hear the story and give their support to Rosa. Eventually, the cashier had to admit her wrongdoing and apologize, which was satisfactory for Rosa as she could have filed a case of racism with the police.

In the case of another research participant, Silvio Humberto—an economist who worked with the government and ran his own NGO—his education, position, and considerable social connections did not protect him from racism. On an evening of celebration, after the successful defence of his doctoral dissertation in Economics at the Federal University of Bahia, he went for dinner with his friends and family to an upscale restaurant in the city centre. He explained that his group was barred from entering the restaurant after a heated exchange with the manager about finding seating inside. He filed a case of racism against the restaurant.

Despite these very public racial incidents, however, race is not commonly thought of as the primary source of stigma, nor is it considered to be a "real" barrier to social mobility for many research participants. For example, Túlio Santana once related a story about his father who was getting off a train one day and was approached by an old, white woman who disparagingly referred to him as a monkey. He brushed it off as ignorance and lack of education—and taught other members in the family to do the same if they were confronted with a similar situation. Others denied experiencing racism or were of the view that racism only affected poorer blacks, not blacks in a better-off position. For example, Jorge Carvalho described his life growing up in a small town in the interior of Bahia as devoid of racism as most people there were the descendants of enslaved people. He argued that he did not experience directly the influence of white people in everyday life until he migrated to Salvador and lived in other large Brazilian cities. In the town where he grew up, he said that some black people had power and were positive role models in his community, so he did not feel out of place and developed a strong black identity. He explained that he did not recognize he was poor and black until he migrated to Salvador. Although he acknowledged that as a young man making life in Salvador, he encountered some social exclusion connected to his colour and background, this usually faded away when people realized he was a man of action, hardworking, and intelligent. As a more mature man with a successful career, he founded the Association of Black Businessmen in Salvador to address the lack of mentorship and opportunities for young black people:

> I created this organization because of the necessity for discussing this issue of financial exclusion between us, because no-one else is going to do it for us. The more difficult question that the Black Movement needs to address is economic access, as we do not have the knowledge to be part of it since most people are outside of the educational system and outside of power. We need to organize ourselves and take advantage of opportunities in the cosmetics and food industries. We need to create our own support systems. (Interview, September 10, 2006)

Other research participants, such as Bruno Teixeira—a single, gay man in his 50s who lived in Rio Vermelho—would often say that he had never experienced any kind of racial discrimination, much to the annoyance of his nieces—Sandra, Anabela, and Anita—who described incidents of racism that they had experienced in public places like shopping malls, where white shop owners and attendants would closely observe their movements. His elder sister, Helena Teixeira, remarked that it was more the case that they were not aware of racism until confronted with it. For example, as a child, she did not notice any difference between herself and her white classmates or neighbours. But one day she was forced to confront the fact that she was viewed as different when a white playmate came over from school. After glancing at the piano, the chandeliers, and luxury carpets in the living room, her young friend exclaimed, "My goodness, you have a house like a white person!" At the time, Helena thought, "What is a black person's house supposed to look like in that case?"

Helena's daughter, Anabela, reported subtle incidences of racism in the personal sphere of family life. For example, she referred to a relative who kept asking her to groom and comb her hair in a respectable manner. She considers this racism, because she thinks that it is about a lack of acceptance for natural, curly hair because of its associations with blackness, as opposed to straighter, wavy hair that is associated with feminine ideals of beauty.

Another research participant, Nubia—a woman in her late 40s who owned two successful vegetarian restaurants—discussed the death of her father as a child and how this worsened the economic situation of her immediate family, whose members lived in a rural town in Pernambuco. Her mother subsequently sent most of her children to live with better-off relatives. She went to live in the house of her aunt in a rural town in Bahia, carrying out unpaid domestic work looking after her aunt's children, cleaning and cooking for the household.[23] She attended school, but not the private schools of her cousins. This unequal treatment created a desire in her to escape domestic servitude: she put herself through school on money she earned from selling sandwiches and sweets, later working in a shop, and then taking a job as a kindergarten art teacher as a young woman. By this time, her aunt had died, so she moved to Salvador where

[23] The situation of poorer, Afro-Brazilian children performing unpaid, caring work in the households of better-off relatives has been noted by other researchers, for example: Twine (1998) and Hordge-Freeman (2015).

she met and married and then later divorced a black man. She described a difficult relationship with him and with other black men she had intimate relationships with, pointing to issues of black male self-esteem and problematic attitudes towards her skin colour (she was lighter-skinned) and gender, especially as she began to achieve financial success and began living in higher status residential neighbourhoods. Therefore, she resolved to only date white men of a higher socioeconomic status—but her later marriage to a white man had its own challenges. For example, he displayed embarrassment to introduce her to his white friends and to take her to certain places in public—and he also had issues with her business achievements.

To sum up, participants were able to point to incidents of racism in everyday life, though less so in terms of recognizing hidden structural forms of racism, though this was changing. It was clear that noticing racial difference (or ignoring it) can be invested with a host of meanings. It can be personally risky; it can be taken as a sign of one's lack of education or refinement; it can be seen as unpatriotic and an insult to the nation; and it is increasingly monitored by the state and subject to legal action. Even asking questions about personal racial identification can be risky, as it was under the Vargas military dictatorship. I discovered this to be the case when I tried to have a conversation about racial identity with one of my respondents at a party. Speaking in hushed tones, she expressed discomfort at speaking publicly about race and pointed me to the 1988 constitutional definition of racial discrimination.

In this chapter, I laid the groundwork for understanding the impact of larger political and social transformations which shaped the racial and class projects of different historical periods. I attempted to give a sense of historically important pathways of mobility, highlighting how the 1950s ushered in a new period of economic and social transformations important to the participation of Afro-Bahians in the Brazilian economy and society. I also introduced the reader to some of my research participants and gave a sense of their experiences with racism and living in an unequal city. Their struggles for status are significant because it is linked to questions of citizenship and the fight against deeply ingrained cultural notions in which only white people are accorded full humanity (Alvez 2014; Alvez and Vargas 2015). While Afro-Bahians sought to improve their lives, the strategies they used were not necessarily different from those of other Brazilians—but they involved different implications and were shaped by different dynamics, as the following chapters will continue to show.

References

Agier, M. 1994. 'Classe' ou 'Raca?' Socialização, Trabalho e Identidades Opcionais. *Bahia: Análise e Dados. Salvador* 3 (4): 7–13.

Alberto, P. 2011. *Terms of Inclusion: Black Intellectuals in the Twentieth-Century Brazil.* Chapel Hill: University of North Carolina Press.

Alvez, J.A. 2014. Neither Humans nor Rights: Some Notes on the Double Negation of Black Life in Brazil. *Journal of Black Studies* 45 (2): 143–162.

Alvez, J.A., and J.C. Vargas. 2017. On Deaf Ears: Antiblack Police Terror, Multiracial Protest and White Loyalty to the State. In *Identities: Global Studies in Culture and Power* 24 (3): 254–274.

Ansell, A., 2014. *Zero Hunger: Political Culture and AntiPoverty in Northeast Brazil.* Chapel Hill: University of North Carolina Press.

Bacelar, J. 1999. Black Brazil. In *Black Brazil: Culture, Identity, and Social Mobilization*, ed. L. Crook and R. Johnson. Los Angeles: UCLA Latin American Center Publications, University of California.

———. 2001. *A Hierarquia das Raças: Negros e Brancos em Salvador.* Rio de Janeiro: Pallas.

Besse, S.K. 1996. *Restructuring Patriarchy: The Modernization of Gender Inequality in Brazil, 1914–1940.* Chapel Hill: University of North Carolina Press.

Borges, D.E. 1992. *The Family in Bahia, Brazil, 1870–1945* [Electronic Resource]. Stanford: Stanford University Press.

Butler, J. 1998. *Freedoms Given, Freedoms Won: Afro-Brazilians in Post-Abolition São Paulo and Salvador.* New Brunswick: Rutgers University Press.

Cardoso, F.H., and Octavio Ianni. 1960. *Cor e Mobilidade Social em Florianópolis: Aspectos das Relações Entre Negros e Brancos numa Comunidade do Brasil Meridional.* Vol. 307 [Coleção Brasiliana). São Paulo: Cia Editora Nacional.

Cicalo, A. 2012. *Urban Encounters: Affirmative Action and Black Identities in Brazil.* New York: Palgrave Macmillan.

Corrêa, M. 1993. Repensando a Família Patriarcal Brasileira. In *Colcha de Retalhos: Estudos Sobre a Família no Brasil*, ed. M. Corrêa. Editora da Unicamp: Campinas.

Corrigan, G. 2015. Is Brazil Making Progress on Inequality? World Economic Forum News Bulletin. https://www.weforum.org/agenda/2015/09/is-brazil-making-progress-on-inequality/. Accessed 30 July 2021.

Costa-Pinto, Luiz. 1998 (1953). *O Negro no Rio de Janeiro: Relações de Raça Numa Sociedade em Mudança.* Rio de Janeiro: UFRJ.

Dávila, J. 2006. *Diploma de Brancura: Politico Social e Racial no Brasil, 1917–1945.* São Paulo: Editora UNESP.

———. 2012. Brazilian Racial Relations: A Changing Context. *The Latin Americanist* 56 (4): 1–10.

De Almeida, P.H. 2006. A Economia de Salvador e a Formação de sua Região Metropolitana. In *In Como Anda Salvador e Sua Região Metropolitana*, ed. Inaiá Carvalho and Gilberto Corsa Perreira. Salvador: Editora Universidade Federal da Bahia.

De Azevedo, T. [1953] 1996. *As Elites de Cor numa Cidade Brasileira*. Salvador: Editora da Universidade Federal da Bahia.

De Queirós Mattoso, K. 1992. *Bahia: Século XIX: Uma Província no Imperio*. Rio de Janeiro: Editora Nova Fronteira.

De Santana Pinho, P. 2008. African-American Roots Tourism in Brazil. *Latin American Perspectives* 35 (3): 70–86.

———. 2010. *Mama Africa: Reinventing Blackness in Bahia*. Durham: Duke University Press.

De Souza, V.S. 1979. Patterns of Urban Growth in Brazil's Southeast and Northeast Regions and Economic Development: The 1940–1970 Period. Masters' Thesis, The Victoria University of Manchester.

Dixon, K. 2016. *Afro-Politics and Civil Society in Salvador da Bahia, Brazil*. Gainesville: University Press of Florida.

Do Valle Silva, N. 1985. Updating the Cost of Not Being White in Brazil. In *Race, Class and Power in Brazil*, ed. Pierre Michel Fontaine, 42–55. Los Angeles: UCLA Center for Afro-American Studies.

Edmonds, E.E. 2010. *Pretty Modern: Beauty, Sex, and Plastic Surgery in Brazil*. Durham: Duke University Press.

Fernandes, F. 1969. *The Negro in Brazilian Society*. New York: Columbia University Press.

Ferreira, F.H.G., Carlos E. Velez, and Ricardo Paes de Barros. 2004. Inequality and Economic Development in Brazil. World Bank Publications. https://elibrary.worldbank.org/doi/abs/10.1596/0-8213-5880-4. Accessed 30 July 2021.

Figueiredo, A. 2003. A Classe Media não vai ao Paraíso: Trajetórias, Perfis e Negritude entre Os Empresarios Negros. PhD dissertation. Rio de Janeiro: Universidade do Rio de Janeiro.

Frazier, F. 1942. The Negro Family in Bahia, Brazil. *American Sociological Review* 7 (4): 464–478.

Freyre, G., and Samuel Putnam. 1986. *The Masters and the Slaves: A Study in the Development of Brazilian Civilization*. Oakland, California: University of California Press.

Gledhill, J. and Hita, M.G. 2009. *New Actors, New Political Spaces, Same Divided City? Reflections on Poverty and the Politics of Urban Development in Salvador, Bahia*. BWPI Working Paper No. 102. Brooks World Poverty Institute. https://papers.ssrn.com/sol3/papers.cfm?abstract_id=1544283. Accessed 5 Aug 2021.

Goldschmidt, W. 1950. Social Class in America: A Critical Review. *American Anthropologist* 52: 483–498.

Graden, D.T. 2006. *From Slavery to Freedom in Brazil: Bahia 1835–1900.* Albuquerque: University of New Mexico.

Guimarães, A.S.A. 1982. A Formação e a Crise da Hegemonia Burguesa na Bahia – 1930 a 1964. Masters Dissertation. Salvador: Universidade Federal da Bahia.

———. 1987. Estructura e Formação das Classes Sociais na Bahia. *Novos Estudos, CEBRAP* 18 (2): 57–69.

Guimarães, A.S.A., Michael, A., and Nadja, A.C. 1995. *Imagens e Identidades do Trabalho.* São Paulo: Editora Hucitec.

Harding, R.E. 2000. *A Refuge in Thunder: Candomblé and Alternative Spaces of Blackness.* Bloomington/Indianapolis: Indiana University Press.

Hasenbalg, C.A. 1979. *Discriminação e Desigualdades Raciais no Brasil.* Rio de Janeiro: Graal.

Herskovits, M.J. 1943. The Negro in Bahia, Brazil: A Problem in Method. *American Sociological Review* 8 (4): 394–404.

Hita, M.G. 2014. *A Casa das Mulheres N'Outro Terreiro: Famílias Matriarcais em Salvador-Bahia.* Salvador: Editora da Universidade Federal da Bahia.

Hordge-Freeman, E. 2015. *The Colour of Love: Racial Features, Stigma and Socialization in Black Brazilian Families.* Austin: University of Texas Press.

Ianni, O. 1966. *Raças e Classes Sociais no Brasil.* Rio de Janeiro: Civilização Brasileira.

Ickes, S. 2013. *African-Brazilian Culture and Regional Identity in Bahia, Brazil.* Gainesville: University Press of Florida.

Ivo, A.B.L. 2008. *Viver Por Um Fio: Pobreza e Politica Social.* São Paulo and Salvador: Annablume and CRH/UFBA.

Jaffe, R. 2016. *Concrete Jungles: Urban Pollution and the Politics of Difference in the Caribbean.* New York: Oxford University Press.

Landes, R. 1947. *The City of Women.* Albuquerque: University of New Mexico Press.

Lima, M. 2012. Affirmative action in Brazil: Challenges for inclusion. *Revista: Harvard Review of Latin America.* https://archive.revista.drclas.harvard.edu/book/affirmative-action-brazil. Accessed 5 Aug 2021.

Machado Borges, T. 2003. Only for You!: Brazilians and Telenovela Flow. PhD dissertation, Department of Social Anthropology, Stockholm University.

Machado de Assis, J. M., and H. Caldwell. 1987 (Originally Published in 1876). *Helena: A Novel by Machado de Assis.* Oakland: University of California Press.

Mahony, M.A. 1997. Afro-Brazilians, Land Reform, and the Question of Social Mobility in Southern Bahia, 1880-1920. *Luso-Brazilian Review* 34 (2): 59–79.

Martins, Silva, S. da Silva, C.A. Medeiros, and E.L. Nascimento. 2004. Paving Paradise: The Road from Racial Democracy to Affirmative Action in Brazil. *Journal of Black Studies* 34 (4): 787–816.

Mayblin, M. 2010. *Gender, Catholicism, and Morality in Brazil: Virtuous Husbands, Powerful Wives.* New York: Palgrave Macmillan.

McCallum, C. 2005. Racialized Bodies, Naturalized Classes: Moving Through the City of Salvador da Bahia. *American Ethnologist* 32 (1): 100–117.

———. 2007. Women Out of Place: A Micro-Historical Perspective on the Black Feminist Movement in Salvador da Bahia, Brazil. *Journal of Latin American Studies* 39 (1): 55–80.

Medeiros, M.A. 2018. *Marriage, Divorce, and Distress in Northeast Brazil: Black Women's Perspectives on Love, Respect, and Kinship*. New Brunswick: Rutgers University Press.

Mendes, M. 2015. *Inequality, Democracy and Growth in Brazil: A Country at the Cross-roads of Economic Development*. London: Academic Press.

Muniz de Albuquerque Jr., D. 1999. *A Invenção do Nordeste e Outras Artes*. São Paulo: Cortez Editora.

Neri, M. 2011. *Desigualdade de Renda na Década*. Report, Fundação Getulio Varas/Centro de Políticas Sociais, Rio de Janeiro.

Neto, P.F.D. 2006. *Tradição, Autocracia e Carisma: A Política de Antonio Carlos Magalhaes na Modernização da Bahia (1954–1974)*. Belo Horizonte: Editora Universidade Federal de Minas Gerais.

Nishida, M. 1993. Manumission and Ethnicity in Urban Slavery: Salvador, Brazil, 1808–1888. *Hispanic American Review* 73 (3): 361–391.

Nobles, M. 2000. *Shades of Citizenship: Race and Census in Modern Politics*. Stanford: Stanford University Press.

Nogueira, O. 1985. *Tanto Preto Quanto Branco: Estudos de Relações Raciais*. São Paulo: T.A. Queiroz.

O'Dougherty, M. 2002. *Consumption Intensified: The Politics of Middle Class Daily Life in Brazil*. Durham: Duke University Press.

Otovo, O.T. 2016. *Progressive Mothers, Better Babies: Race, Health, and the State in Brazil, 1850–1945*. Austin: University of Texas Press.

Parker, R.G. 1991. *Bodies, Pleasures, and Passions: Sexual Culture in Contemporary Brazil*. Nashville: Vanderbilt University Press.

Pereira, A.W. 1997. *The End of Peasantry: The Rural Labor Movement in Northeast Brazil*. Pittsburgh: University of Pittsburgh Press.

Perry, K.K.Y. 2013. *Black Women Against the Land Grab: The Fight for Racial Justice in Brazil*. Minneapolis and London: University of Minnesota Press.

Pierson, D. 1942. *Negros in Brazil: A Study of Race Contact at Bahia*. Chicago, Illinois: The University of Chicago Press.

Rebhun, L.A. 1999. *The Heart is Unknown Country: Love in the Changing Economy of Northeast Brazil*. Palo Alto, California: Stanford University Press.

Reis, J.J., and H. Sabrina Hill. 2003. *Death is a Festival: Funeral Rites and Rebellion in Nineteenth Century Brazil*. Chapel Hill: University of North Carolina Press.

Rocha, L. 2010. Research (ing/in) State Genocide: Toward an Activist and Black Diasporic Feminist Approach. MA Report: University of Texas at Austin.

Rodrigues, N. 1935. *L'Animismo Fetichista dos Negros Bahianos*. Rio de Janeiro: Civilização Brasileira.
Romo, A.A. 2010. *Brazil's Living Museum: Race, Reform, and Tradition in Bahia*. Chapel Hill: University of North Carolina Press.
Sanabria, E. 2016. *Plastic Bodies: Sex Hormones and Menstrual Suppression in Brazil*. Durham: Duke University Press.
Santos, S.B. 2008. Brazilian Black Women's NGO's and Their Struggles in the Areas of Sexual and Reproductive Health. PhD dissertation, University of Texas at Austin.
Schwartz, S.B. 1974. The Manumission of Slaves in Colonial Brazil: Bahia, 1684–1745. *The Hispanic American Historical Review* 54 (4): 603–635.
Selka, S. 2007. *Religion and the Politics of Ethnic Identity in Bahia, Brazil*. Gainesville: University Press of Florida.
Skidmore, T.E. 1974. *Black into White: Race and Nationality in Brazilian Thought*. New York: Oxford University Press.
Smith, C.A. 2016. *Afro-Paradise: Blackness, Violence, and Performance in Brazil*. Urbana: University of Illinois Press.
Souza, A.G. 2000. *Limites do Habitar: Segregação e Exclusão na Configuração na Contemporânea de Salvador e Perspectivos no Final do Século XX*. Salvador: Editora da Universidade Federal da Bahia.
Telles, E.E. 2004. *Race in Another America: The Significance of Skin Color in Brazil*. Princeton: Princeton University Press.
The Institute of Brazilian Geography and Statistics. 2010. https://cidades.ibge.gov.br/brasil/ba/salvador/panorama. Accessed 5Aug 2021.
Twine, F.W. 1998. *Racism in a Racial Democracy: The Maintenance of White Supremacy in Brazil*. New Brunswick: Rutgers University Press.
Vargas, J.H.C. 2016. Black Disidentification: The 2013 Protests, Rolezinhos, and Racial Antagonism in Post-Lula Brazil. *Critical Sociology* 42 (4–5): 551–565.
Warner, W.L. 1959. *The Living and the Dead. The Symbolic Life of Americans*. New Haven: Yale University Press.
Warner, W.L., and Lunt, P.S. 1941. *The Social Life of a Modern Community*. New Haven: Yale University Press.
———. 1942. *The Social Status of a Modern Community*. New Haven: Yale University Press.

CHAPTER 3

Racial Discourses and Identities

I say that the Black masses in Brazil have only one option: to disappear. Whether it be through compulsory miscegenation or: when they escape this, through direct elimination—death pure and simple. This has been going on for four centuries. (Abdias do Nacsimento [1979] 1989, 7)

Inevitably, many interpretations of race, many racial discourses, exists at any given time. The political character of the racial formation process stems from this: elites, popular movements, state agencies, cultural and religious organizations, and intellectuals of all types develop racial projects, which interpret and reinterpret the meaning of race. (Howard Winant 1994, 24)

In Bahia, a small but powerful Euro-oriented elite have long wielded enormous political, economic, and social control over a state where the African-descendant population is in the majority. For years, a group of white right-wing male populists dominated Bahian politics, famously led by the late Antonio Carlos Magalhães, who was governor of Bahia three times and represented Bahia in the Senate of Brazil three times.[1] Magalhães

[1] See Tom Phillips, "Antonio Carlos Magalhaes." *The Guardian*, 23 July 2007. https://www.theguardian.com/news/2007/jul/23/guardianobituaries.brazil

There are some exceptions: the Left-Wing Lidice da Mata, a member of the Brazilian Socialist Party, represented Bahia in the Federal Senate from 2011 to 2019. Previously, she was mayor of Salvador from 1993 to 1997, but her ability to govern was tightly restricted by lack of funds. Being unable to maintain basic services, she suffered constant criticism in the press.

D. J. Gordon, *Blackness and Social Mobility in Brazil*,
https://doi.org/10.1007/978-3-030-90765-5_3

was one of Brazil's most powerful and influential politicians for at least 50 years, serving as a minister for communications, as leader of the Liberal Front Party, and as president of the Federal Senate.[2] Despite the emergence of a vibrant black civil society (Dixon 2016) and the increased celebration of Afro-Brazilian cultural forms (De Santana Pinho 2010), it is clear that opposition to the dominant political and economic group and to glaring structures of inequality in Salvador are yet unable to find a clear way to power. Thus, it is imperative that we understand the ideational aspects of race if we are to understand the structures of power at work and learn the varied ways in which anti-racist organizations will have to operate to dismantle racism, both locally and globally. How are racial ideologies and discourses reproduced and/or transformed over time? How are racialized ideologies and hierarchies located within transnational fields? How do they shape the vulnerabilities and exclusions of racialized and gendered populations? How do Afro-Brazilians participate, reproduce, or reject racialized hierarchies, discourses, and practices in their daily lives? These are some of the questions explored in this chapter.

In their book on racial formation in the United States, Michael Omi and Howard Winant argue that becoming a citizen of this society involved a process of learning to "see" race—to ascribe social meanings and qualities to otherwise meaningless biological features (1994, 52). The authors explore the historical development of race as a dynamic and fluid social construct, showing how racial formation occurs at two distinct levels: the macro-level of social institutions, social structures, common ideologies, and political projects as well as the micro-level of social relations—the ways in which we understand ourselves and interact with others. Indeed, our experiences are processed through practice, discourse, and interpretation (Skeggs 1997). In the following sections, I will move between these different levels, exploring global racial formations, national ideologies, and everyday interpersonal interaction in Salvador.

Comparative Racial Formations

In the Americas, the violent legacies of plantation slavery and colonialism created similar racial formations in places like the Brazilian Northeast and the Caribbean (Beckford 1972; Wagley 1957). In these contexts, whiteness was a transnational force, part of an interconnected global system that

[2] Ibid.

historically ranked cultures and societies in terms of their superiority/inferiority in relation to Europe, in both racial and cultural terms. Right across Latin America, discourses and practices of *embranquecimento* ("whitening")—and the concomitant practice of *mestiçagem* (race mixture)—have been described by numerous scholars as being integral to regional white supremacy ideologies (Skidmore 1974; Wright 1990; Wade 2007). It is therefore not surprising that with limited opportunities for upward mobility, strategies of "whitening" (such as marrying into white families or adopting lifestyles associated with the white elite) represented a means by which African descendants across the diaspora could improve their lives and social position. As the Martinique-born Franz Fanon suggests in *White Skin, Black Masks* ([1952] 2008), surviving white supremacist systems has often entailed wearing masks. Masks are symbolic coverings that attempt to shield the wearer from an oppressive gaze. However, in doing so, mask wearers run the risk of internalizing the distorted image that is projected unto us, a kind of symbolic violence.[3]

In both Brazil and the Caribbean, however, the creation of national ideologies of race mixture, the impact of social movements such as the Black Conscious Movement, the dynamism of civil society organizations, the influence of African-derived religions, and the implementation of multicultural policies have challenged the strong "white bias" in the region's plantation societies.[4] Nationalizing discourses of *mestizaje* (racial mixing) within the Latin American and Caribbean context are seen largely as positive processes, yet they nevertheless are unsettled, relational projects where "people are not white or black, but rather, they are whiter than or darker than others" (Moreno Figueroa 2013, 139). Celebrations of *mestizaje* are common to the region; however, this often obscures hierarchies of race and class that reinforce the dominant value of whiteness, which is part of the region's historical legacy.

[3] Born under French colonial rule on the Caribbean island of Martinique, Frantz Fanon (1925–1961) was one of the most influential writers in Black Atlantic theory during a period of anti-colonial struggle. His work drew from a mix of poetry, psychology, philosophy, and political theory, and its influence across the global South has been powerful and enduring. In his lifetime, he published two key works which were translated into English, *Black Skin, White Masks* in 1952 and *The Wretched of the Earth*, in 1961.

[4] The term "white bias" comes from Fernando Henriques, *Family and Colour in Jamaica*. However, note that in demographic terms, Brazil has a larger white population than many countries in the Caribbean.

Latin American and Caribbean racial formations have been compared to racial formations elsewhere, such as the segregationist southern United States and Apartheid South Africa. In South Africa, there was minimal experience with slavery except in the Cape. Competition and conflict between colonial powers in South Africa led to less stability than that existed in Brazil where consistent maintenance of Portuguese rule helped to reduce internal conflict (Marx 1998, 45). Unlike South Africa or the segregationist American South, racial categories were not imposed legally in Brazil. This is not a trivial distinction; when racial discrimination is mandated by law, it operates in a more rigid and inflexible way than in societies in which it is informal and at the discretion of individuals. At the same time, the informality of racial discrimination has made racism difficult to name in Brazil.[5]

While racial formations in countries like Brazil and the Caribbean were more flexible and allowed for boundary crossing, South African racial categories were more rigid. However, in Brazil, it has not been uncommon for contradictory discourses about race to co-exist—such as an identification with blackness while simultaneously distancing oneself from the stigmatizing meanings of blackness. In addition, there is continued silence around Brazil's violent past and present. Brazilian black activists remind us that police or state violence against blacks in Brazil is an ongoing phenomenon (Hafiz 2016). Citing data from the Instituto de Pesquisa Econômica Aplicada (Institute of Applied Economic Research), Perry (2019, 159) has stated that between 1980 and 2014, a record one million black people (approximately 29,000 per year) were killed. Furthermore, black men and women—who make up roughly half of the population in Brazil—comprised 71 per cent of the 318,000 homicide victims in Brazil between 2005 and 2015 (Perry 2019, 159). Thus, while explicit ideas about race are generally not expressed in the context of Brazil's racial formation, anti-blackness and the sponsorship of racial violence are ever present.

[5] Earlier versions of these comparative ideas about race were developed in previous publications by the author. Please see: "Body Beautiful: Comparative Meanings of Beauty in South Africa, Brazil and Jamaica," In *Beauty and the Norm: Debating Standardization of Bodily Appearance*, edited by Sarah Böllinger, Claudia Liebelt, and Ulf Vierk (2018, Palgrave Macmillan); and "Crossing Borders, Blurring Boundaries: Comparative Meanings of Beauty in Brazil, South Africa and Jamaica," Special Issue on Beauty and Sexuality (2016, *Ideaz* 14: 101–132.).

Race and Progress in Brazil

The preoccupation with race in Brazil is rooted in the country's history and is discussed in greater detail elsewhere (Skidmore 1999; Schwarcz 1993). The period between 1870 and 1910 was an important one historically for Brazil, as it ushered in the end of slavery, the transition from a Brazilian empire to the country's first republic, the beginnings of industrialization, and efforts by the elite to improve the image of Brazil at home and abroad in both racial and cultural terms. As the end of slavery neared, politicians, planters, intellectuals, scientists, and physicians debated Brazil's future and transition to wage labour. More specifically, the elite harboured racist attitudes towards the non-white labour force, who they saw as physically inferior and incapable of serious work habits (Skidmore 1999, 71–73).

Theories of scientific racism prominent in Euro-America in the nineteenth century influenced ideas about racial mixture being the cause of degeneration and backwardness in Brazil. Political and social elites saw their country as composed of a largely unhealthy, unstable, and even dangerous populace whose members were perceived as not being capable of fuelling the development of a modern, competitive nation. Many solutions to this situation were discussed, including a plan to gradually replace Brazil's large non-white working population with white immigrants, as whiteness was desired for its civilizational potential. Brazilian elites began putting their trust in miscegenation—a process that horrified white Americans in the southern United States who had just implemented "Jim Crow" laws to keep the races physically separated in public. Rather, Brazil's elite created an ideology of whiteness which depended on the process of racial mixing. In the words of João Batista Lacerda, a leading doctor and anthropologist who spoke at the First Universal Race Congress in London in 1911, "in the course of another century, the mixed bloods will have disappeared from Brazil. This will coincide with the extinction of the black race from our midst" (cited in Skidmore 1999, 78). By 1860, plans to encourage European immigration had developed into pilot projects being tested out across the country. While laws were not used to keep Afro-Brazilians "in their place," colour barriers were observed, for example, in the Ministry of Foreign Affairs, the officer corps of the navy, and the higher levels of the Catholic Church (Skidmore 1999, 79).

Theories of eugenics and social Darwinism seemed to offer some hope for "improving" Brazil's racial stock. The idea was to intervene in the reproductive processes of populations by identifying "abnormal" physical characteristics and restricting their propagation. A version of this theory was to become popular in Brazil and focused on a new social politics of intervention (Stephan 1991; Dávila 2006). For example, part of improving Brazil's image abroad included efforts to Europeanize the physical appearance of some of its cities and to improve public health (Skidmore 1999). Thus, race became linked to the very future of a people and a nation: Brazil's population could be "improved" through social engineering and various techniques of physical and social whitening. This ideology was embraced by Brazilian elites, as the country was promoted as a mixed-race nation in the process of whitening, symbolizing what the nation would become in the future (Twine 1998).[6]

However, Brazilian ideas about race remained contradictory. No doubt, ideas about *mestiço* degeneracy were fed by fears of the upward mobility of their kind into the bourgeois classes. For example, it was known that persons professing to be whites also included persons "tainted" by mixed blood (Pierson 1942, 128–129). The supposedly unstable and violent character of *mestiços* was the subject of medical, scientific, psychological, and intellectual studies. In *The Spectacle of the Races* (1999), Brazilian historian and anthropologist Lilia Moritz Schwarcz has analysed the competing but complementary schools of thought emanating from the Bahia and Rio de Janeiro Schools of Medicine at the turn of the century, which played a role in articulating the medical interest in bodies as a means of perfecting the nation. The Bahia School of Medicine promoted its own version of ideas about racial mixing, which was thought to explain criminality, degeneracy, and certain kinds of illnesses. Physicians were also deeply concerned and engaged in debates about race, population improvement, and progress. For example, they proposed that the state could invest in its "human capital" through maternal and child health and welfare programmes (Otovo 2016). It was believed that slavery and its social institutions—especially the personalized practices of midwives, wet-nurses, and nursemaids—had created distortions in the society that could be rectified by paying attention to the education and health of the popular classes, whose labour was needed to create a progressive nation (Otovo 2016, 10). Drawing from international debates about maternalism, Brazilian policy makers targeted north-eastern states with a higher percentage of black residents. Between the years 1850 and 1945, maternalism was institutionalized into policies in specific ways in Bahia, targeting poor black Bahian women, their families, and their

[6] These points were developed in a previous publication by the author. Please see: "A Beleza abre Portas: Beauty and the Racialized Body among Black Middle-Class Women in Salvador, Brazil" (July 17, 2013, *Feminist Theory* 14: 203–218).

child-rearing practices, with a view to modernizing the masses. Black Bahian women were therefore being incorporated into the state in specific ways that reinforced their place at the bottom of Brazilian society, though they were also finding much needed medical assistance and social aid (Otovo 2016, 2).

However, it was clear that despite these policies of population improvement, Brazil remained far from white, especially in the Northeast as most of the white immigrants went to the Southeast and South of Brazil. This implied that states like Bahia, with majority black and mixed-race populations, were destined to be undeveloped and uncivilized. An alternative response to this situation was to find a positive element in Brazilian racial mixture. Early ideas about valuing Brazilian racial mixture can probably be traced back to the abolitionist period in Brazil—however, Gilberto Freyre is best known for fully articulating these thoughts, as noted previously. From here on, the ideal for the nation of Brazil was to be based on a celebration of mixture, represented by Brazilians' cordial, personal, and flexible character, as in Sergio Buarque De Holanda's ([1936] 2006) term, *o homem cordial.* Gilberto Freyre's work shifted the way Brazilian social and political thought would view Africans and their descendants. With the focus now on culture and African people's contribution to Brazilian civilization, the older racialist discourse of science was challenged. This new perspective was expressed in the regionalist literature of the time and in the emerging cultural industry, as a modern Brazilian nation was crafted—one that had overcome racialism and which could celebrate the cultural heritage of Brazilian blacks, *mulatos* (mixed-race Brazilians) and *caboclos* (mixed with indigenous heritage) in a racial democracy.

As Roth-Gordon (2017, 21) has noted, both of Brazil's twentieth-century authoritarian regimes were heavily invested in the idea of *mestiçagem* (racial and cultural mixture). President Getúlio Vargas (dictator from 1930 to 1945, democratically elected from 1951 to 1954) actively promoted *brasilidade*—feelings of Brazilian nationalism, resting on notions of the "racially harmonious Brazilian national family" (Davis 1999, 2). The idea that Brazil is a racially tolerant nation gained an even stronger foothold during Brazil's military dictatorship from 1964 to 1985, despite the findings of an internationally publicized UNESCO study on race in the early 1950s that pointed to both subtle and statistical evidence of racial inequality and race prejudice in Brazil (Roth-Gordon 2017, 22). The propaganda around Brazil's racial democracy and racial tolerance was not intended merely for a Brazilian audience—the pride in a racially mixed Brazilian identity and in a country that had supposedly achieved racial harmony was constantly evoked in relation to North America's reputation as a land of racial hatred and division (Seigel 2009, 217). These sentiments have hardly disappeared in more recent times, as the idea of a "Brazilian national family," inclusive of African, European, and indigenous peoples, was broadcast to the world during the World Cup in 2014 in Brazil as three children (one white, one black, and one indigenous) were chosen to release white doves in the opening ceremony. More recently, President Bolsonaro has revived the idea of Brazilian diversity and the non-existence of racial division in Brazil, emphasizing that everyone is Brazilian, and therefore mixed race.[7]

Cultural Politics and Racial Discourse in Bahia

Associated closely with the racial democracy ideology is a discourse of cultural democracy, where each culture was supposedly given its place in the making of the Brazilian nation. In Bahia—the centre of African cultural survivals and adaptations—this emphasis on culture took on greater proportions than in other Brazilian states. Hanchard (1994) has argued that an Afro-Brazilian "culturalism," based on the affirmation of African survivals and identities, has been subject to co-optation by the government and the white establishment. Some scholars argue that official willingness to promote and celebrate the African strain in Brazilian national character and culture, which dates back to the 1930s, has served to distract attention from the economic and social plight of Afro-Brazilians and inhibit protest against poverty, discrimination, and police brutality (Dixon 2016; Roth-Gordon 2017; Smith 2016). For example, during my fieldwork in Brazil, an open-air exhibition by Brazilian artist, photographer, and cultural producer, Sergio Guerra, was staged in different parts of the city. The exhibition, called "Salvador NegroAmor," took place between January 8 and February 16, 2007, with the objective of initiating "a grand movement of valorization of the Afro-descendant people of Salvador."[8] It was considered the largest photography exhibition of its kind, showing 1501 photographs and covering an area of 10,000 square metres of the city, utilizing street light posts, parks, buildings, plazas, transportation centres, and markets. The effect was to highlight—and even romanticize—the city's African-

[7] Anthony Faiola and Marina Lopez, "How Jair Bolsonaro Entranced Brazil's minorities—While Also Insulting Them." *Washington Post*, October 24, 2018. https://www.washingtonpost.com/world/the_americas/how-jair-bolsonaro-entranced-brazils-minorities-while-also-insulting-them/2018/10/23/a44485a4-d3b6-11e8-a4db-184311d27129_story.html

[8] See the artist's website, accessed on August 17, 2020, at: http://www.sergioguerra.com/exposicao-detalhe.asp?id=4

descendant population, while neutralizing Salvador's obvious race and class inequalities in which black people are positioned at the bottom of the hierarchy and in peripheral neighbourhoods.

While Bahian elites have begun to celebrate and exalt particular "African" traditions and practices, it is important to note that their attitudes differed in the past. During the late nineteenth century, Butler (1998, 182–183) noted how letters to the editor of local Bahian newspapers revealed the fears of the white elite about "displays of African culture," which they saw as a source of profound embarrassment. To them, these characteristics—which were revealed in full force in popular cultural events such as carnival and annual religious festivities—placed them on a lower level than civilized countries and unfavourably positioned Bahia in relation to the more developed parts of Brazil.

By the 1930s, however, a growing regionalism developed among Bahian elites. They increasingly employed an idiom of race to distinguish themselves from other Brazilians. Among other states, Bahia symbolized *A Mulata Velha,* or the Old Black Mammy, the nursemaid who nurtured and sustained culture for the children of the nation (Borges 1992, 31). This heritage began to be celebrated in cultural productions about the city. For example, the African descent heritage of Bahia was showcased by Bahian elites in their presentation of Afro-Bahian food—*vatapá, acarajé, caruru,* and *frigideira de caranguejos*[9]—to foreigners. Moreover, certain sections of the Bahian elite began to celebrate Candomblé as the prime marker of Bahian cultural heritage—a marked change from their persecution of the religion in previous decades. The celebration of certain aspects of the religion was taken up by artists and intellectuals of the time. The *terreiros* (temples of worship) belonging to the *nagô-jeje* tradition, which tended to stress their faithfulness to African tradition, were often the favoured ones, marginalizing the less acceptable faces of popular Candomblé, which incorporated syncretic practices into their religious ritual, such as *caboclo* (indigenous) deities (Henfrey 1981; Matory 2005). The emphasis on purity, which began to be celebrated by some black Bahians and was later appropriated by some parts of the Brazilian Black Movement, was not really a concern of elites. Elites instead developed an

[9] *Vatapá* is a kind of shrimp or seafood stew cooked in coconut milk: a thick version of this stew is sometimes served with *acarajé*, which is a crispy bean and onion cake typically fried in palm oil. *Caruru* is a dish made with okra, onion, shrimp, palm oil, and toasted nuts (peanut or cashew). It is a food typically served at Afro-Brazilian religious ceremonies and to celebrate ties of family and friendship. *Frigideira de caranguejos* is made with shredded crab meat, garlic, onions, tomatoes, sweet peppers, coconut milk, breadcrumbs, and butter.

attitude of protectionism towards certain aspects of Candomblé religious practice as "cultural patrimony" of Bahia.

A growing tourist and culture industry increasingly promoted a "Bahian cultural heritage" based on the state's large black population. In his book about cultural policies in Bahia, Santos (2005) demonstrates how Vargas' policy of decentralized industrial development was linked to the idea of profiting from local particularities for economic growth. Therefore, it naturally followed that Bahia's history, cultural heritage, and beauty could be promoted as a tourist attraction. Economic interests and policies became sharply oriented towards Bahia's cultural capital and deep roots in African culture. In 1962, President Jânio Quadros took this idea further by incorporating what he thought were black cultural manifestations into official government policy (Santos 2005, 54). This line of action was then developed further by the military dictatorship as aspects of "black" culture and folklore began to receive public resources, especially Capoeira, traditional cuisine, music, and Candomblé.

Currently, Afro-Bahian culture is considered *legal* (cool)—but some argue that it is being manipulated, exploited, and marketed by various stakeholders for different reasons, including black cultural groups and the Euro-Bahian elite (in the form of government institutions, politicians, tourism, and commercial interests) whose members frequently work in partnership with each other (Dixon 2016; De Santana Pinho 2010). Sections of the Bahian elite started investing heavily in the concept of *baianidade* (Afro-Bahian culture), a process referred to by Patricia Pinho as "milking mother Africa" (De Santana Pinho 2010, 10). This is reminiscent of a robust conversation that occurred during my fieldwork in Salvador while having drinks at a trendy bar with some Brazilian friends. They were debating the concept of *baianidade* in relation to a film, *Cidade Baixa*, which had just been released.[10] On one side of the debate, some argued that the film, which was set in Salvador, was very stereotypical and that there was nothing interesting about the storyline of two men falling in love with and competing for the attention of a beautiful prostitute. On the other side of the debate, some argued that the film captured the essence of *baianidade*—a complex of meanings associated with the landscape, linguistic expressions and phrases, the city's Afro-descendant people, the sounds, visual and sensual images of Salvador, imagined by Bahians and others inside and outside of Brazil as a specific, racialized locale in the country.

[10] "Cidade Baixa," or "Lower City," is a Brazilian drama film that was released in 2005, directed by Sérgio Machado and produced by Walter Salles and Mauricio Ramos.

The astute white Bahian politician, ACM, understood the potential appeal of Afro-Bahian culture in Salvador and in Brazil more widely. However, ACM's moves towards the figurative and literal embrace of Afro-Bahian culture took time to evolve towards the forms they took when a modern multicultural politics became the order of the day. Although ACM's early political career was marked by conservatism and claims that Brazil was a Catholic country, his style of government changed over the years to the point of his being surrounded by Afro-Bahian Candomblé priests and priestesses on the campaign trail and patronage of traditional Afro-Catholic religious festivals, such as Lavagem do Bomfim (washing of the steps of the Bomfim church). Thus, ACM was a politician capable of changing with the changing times as new political expressions of blackness later turned him into other directions and the celebration of links with Africa also acquired a value that it had not had in the past.

ACM was remarkably popular in Bahia where he learnt to appropriate and patronize aspects of Afro-Brazilian culture to his advantage, seeing that the electorate was largely African descendant. He played a crucial role in the cultural heritage programme launched in the state of Bahia in the early 1990s. The focus of the state on the culture industry in Bahia has tended to emphasize that Afro-Brazilian forms of folklore are to be supported as truly Brazilian, and not simply African, resulting from Brazil's miscegenation and racial democracy.

The historical city centre of Salvador, Pelourinho, is considered the heartbeat of Afro-Bahian culture. Given Pelourinho's unique architectural and cultural qualities, it was officially recognized as a world heritage site by UNESCO in 1985. In this locale, Afro-Bahian cultural productions and historical memory—central to the articulation of Afro-Brazilian identity more generally—are structurally connected to the political economy of tourism and larger transnational forces of commodification. Impressive colonial architecture dating back to the seventeenth and eighteenth centuries, pebble stone streets, Capoeira demonstrations, *acarajé* stalls, traditional rum shops, paintings displayed on sidewalks depicting traditional black culture and north-eastern landscapes, and upscale stores selling Bahian music, crafts, and photographic images of Pierre Verger (a noted French anthropologist) are presented attractively for consumption. Interspersed with this activity, there are non-governmental organizations and cultural groups that occupy real estate in Pelourinho.

The anthropologist John Collins has written about the transformations and renovations that Pelourinho has undergone since it was declared a heritage site. These processes were driven to a large extent by ACM's focus on developing Salvador as a tourist destination—but it has had deep consequences for the largely poor, black residents who occupied the area. Collins has argued that what happened in Pelourinho illustrated the gap that emerged between officially approved expressions of Bahia's African heritage and the reality of residents' daily lives, who fought to stay in their homes as real estate values skyrocketed (Collins 2004, 2015). These grass-roots land struggles are also occurring to different degrees in various other neighbourhoods across Salvador (see Perry 2013). Figure 3.1 shows a sign on a bar in Pelourinho that symbolizes the Brazilian system of *coronelismo*, or patron-based strong-man politics.

The two images of Bahia mentioned in the previous chapter—"Black" and "Brown" Bahia—compete in representations of the restored heritage site of Pelourinho. On the one hand, it is the centre of Afro-Brazilian cultural production and political organization. The name of the neighbourhood translates into "pillory," which stood for the slab of marble to which slaves were bound, humiliated, and tortured. The remains of this site are in the Largo do Pelourinho, attesting to the recent slave history of Salvador. Furthermore, the internationally acclaimed *bloco* (carnivalesque association) and cultural organization, Olodum, established themselves in this neighbourhood, highlighting African heritage and black pride through music, dance, theatre, and art. It is also the neighbourhood where the first church for blacks was built (Nossa Senhora do Rosário dos Pretos: Our Lady of the Rosary of the Blacks).

On the other hand, Pelourinho is the "cradle of Brazil," or "Brazil's Brown Mother" (Collins 2004). As a heritage site, it produces the effect of mixing: for example, the colonial architecture results from a merging of Portuguese and African elements. Olodum has also been interpreted by some as a *mestiço* organization (mixed racial/cultural origins). When it was initially formed as a break-away group from another Afro-Brazilian cultural organization—Ilê Aiyê—Olodum was at first stereotyped as an undignified *bloco*, especially by those promoting an emphasis on ethnic and cultural purity. When Ilê Aiyê was established in the working-class district of Liberdade in the 1970s, it was associated with the *nagô-jeje* Candomblé tradition and promoted a discourse of cultural purity (Magalhães 2006, 56–57). This meant that only dark-skinned persons could enter the carnival *bloco* parade. This policy has since changed; however, it is still known throughout the city as a distinctively black cultural organization. In the following section, I will explore the discourses and

Fig. 3.1 "ACM: Force of Brazil" sign in the Santo Antonio subdistrict of Pelourinho, Salvador. (Source: John Collins 2015, 87)

narratives of whiteness, blackness, and *mestiçagem* in Brazil so that we can better understand its practical operation on the ground and ultimately for the purposes of reflecting on ways of undoing race-thinking.

Discourses of Whiteness

As several scholars have long shown, the social value given to whiteness does not exist only at the level of discourse, but also of practice, enacted in daily social relations (De Azevedo 1996 [1953]; Twine 1998; Dávila 2006; Roth-Gordon 2017). Whiteness has historically been an extremely desirable characteristic and a strong indicator of national inclusion. At the same time, there has been a shift in racial discourse in recent years, such that less than half of the Brazilian population (48 per cent) classified themselves as white on the national census (IBGE 2012).[11]

[11] IBGE (Instituto Brasileiro de Geografia e Estatística). 2012. Censo Demográfico 2010. Brasilia: IBGE. English Language readers may be interested in the following article by Moriano Castillo, "Minorities now Officially a Majority in Brazil". CNN, June 17, 2011. http://edition.cnn.com/2011/WORLD/americas/06/16/brazil.race/index.html.

Scholars writing on whiteness in Brazil have noted that whiteness is a kind of invisible standard against which other races/colours are measured (Reiter 2010; Dávila 2006; De Santana Pinho 2009; Roth-Gordon 2017). To illustrate this point, I will discuss an exploratory trip I made to Brazil in the summer of 2003, during which I visited Rio de Janeiro. While I was there, I made sure to see the Centro Cultural Banco do Brasil's (Bank of Brazil's Cultural Center) exhibition, "Arte da África." The exhibition was showcasing artefacts from the Ethnological Museum of Berlin. Advertised as part of a series of art exhibitions that focused on the "roots" of Brazilian civilization, the exhibition space was divided into three sections: figurative sculptures, performance culture (including masks and musical instruments), and design.[12] As I moved through the gallery viewing the impressive collection of around 300 art pieces from sub-Saharan Africa, however, I was struck by how the backdrop for all the artefacts—whether painted or draped in cloth—was black, which I understood to be symbolically important, perhaps playing on the idea of Africa as the exotic Dark Continent.

Later in 2005, when I began my fieldwork in Salvador, I returned to Rio de Janeiro and attended the second exhibition in the same series, this time called "Por Ti América," which was intended to give a positive recognition and understanding of indigenous cultures and peoples. I stayed with some white Brazilian friends in the Rio de Janeiro neighbourhood of Botafogo. I asked them—will this series of exhibitions by the Banco do Brasil (Bank of Brazil) explore whiteness? I was met with puzzled looks and silences. Instead, I learned that there had been an exhibition in 2004–2005, which presented archaeological objects associated with early man on Brazilian soil, entitled "Antes—Histórias da Pré-Historia." They did not know of an exhibition in the series that examined the European colonizers. I concluded that there was a deafening silence surrounding discussions on whiteness in Brazil—while the country's non-white populations were exoticized, fetishized, or vilified (e.g. as poor shantytown residents, criminals, or drug dons).

[12]The series was discussed in the catalogue, "Por Ti América: Arte Pré-Colombiana," published by the Centro Cultural Banco do Brasil (2005). In addition, the event was discussed in local newspapers and online forums. For example, see archives of the Correio do Brasil, https://arquivo.correiodobrasil.com.br/arte-da-africa-bate-recorde-de-publico-do-ccbb-rio/

While living in Salvador, I pressed my white Bahian friends about the existence of racism in Brazil. At times they were defensive and insisted that they had black Brazilian friends and that they mingled with everyone, regardless of social background. They would point out that their participation in aspects of Afro-Bahian culture meant that they were not racist. For example, they may eat Afro-Bahian food, listen to Afro-Brazilian musical artists, go to Afro-Brazilian religious ceremonies, or visit Pelourinho, where Afro-Brazilian culture is on display, to socialize. For the most part, they accept that there is racial prejudice and racial inequality in Brazil—but are less likely to acknowledge racism as being embedded in wider power relations in which white people hold a privileged position and where broad social structures and institutions shape Afro-Brazilians' exclusion from citizenship.

However, whiteness is not a stable or straightforward identity in Bahia—nor elsewhere for that matter (Hartigan 2000; Roth-Gordon 2017).[13] The term *branco de Bahia* (Bahian white) suggests that Bahian whites are not from a "pure" racial heritage and have some racially mixed ancestry. For example, Liv Sovik (2009) entitled her study of Bahia *Aqui Ninguem é Branco* (*There Are No Whites Here*). Most light-skinned Bahians do not refer to themselves as white, preferring to describe themselves as *moreno* (or the feminine form *morena*; brown/mixed race) or *moreno claro* (light brown). Indeed, very fair skin can be stigmatized as unattractive, in need of some sun-tanning. At the same time, common phrases such as *vai na sombra* (*walk in the shade*) suggest that one should avoid getting too dark. Thus, the threat of symbolically "blackening" is always there. One could blacken not just in terms of marrying into a black family, but also through inappropriate behaviour (Fischer 2004), by living in a neighbourhood associated with blackness or becoming downwardly mobile. Therefore, race and class do not always map unto each other in expected ways to produce a clear concept of whiteness.

Whiteness is both invisible and visible at different times and in different contexts, and it is shaped by class and gender dimensions. For example, it

[13] In a Special Issue of the journal, *Identities*, Hartigan (2000) explains that whiteness has emerged rapidly as an intense subject of scrutiny for activists and academics. He argues that rather than trying to define whiteness, it is more useful to reflect on its meaning and how it is deployed in specific contexts within a global hierarchy of privilege. Furthermore, whiteness is not just associated with European colonialism and capitalistic domination but also reveals itself in a range of social relationships between people, as well as positions of power that are negotiated in relation to narratives of whiteness (2000, 270).

is implied when Brazilians use terms such as *os ricos* (the rich people), *a classe alta* (the upper class), *os barões* (the barons) *gente bonita* (the beautiful people), or *boa aparência* (good appearance). Furthermore, the very rich in Bahia are hidden as their privileged class status affords them privacy—they live in high security residences, drive in cars with tinted windows, or take helicopters from one city to the next. Occasionally, the gossip pages of the local newspapers will have a story about an upcoming wedding—such as the lavish wedding of an Odebrecht family member which took place during the time of my fieldwork and involved the closing off some streets to vehicular traffic. Apart from these more public showings, to most ordinary Bahians, the lives of the Euro-Bahian elites are invisible.

On the other hand, white or light skin visibly stands for social status in Brazilian society, as part of a "larger economy of signs" that includes bodily features and hair texture (De Santana Pinho 2009, 40). In Brazil's folk taxonomy of colour, there is a hierarchy of descriptive terms relating to the perceived superiority of "European" physical features on one end of the spectrum and supposed inferiority of "African" physical features on the other, which I discuss in more detail in Chap. 7. According to De Santana Pinho (2009), the hyper-valuation of European physical appearance—as evidenced by the dominance of white or light-skinned Brazilians on television screens, billboards, and magazines—is firmly in place in Brazil.

The Brazilian historian Jerry Dávila has underscored that while whiteness is desired, it has always been precarious; Brazilians already thought that they lacked the whiteness they needed to be among the "civilized" nations, and so it was necessary to find ways of creating whiteness, such as marrying into Euro-Brazilian families or adopting the lifestyle and consumption practices of the lighter-skinned elite (Dávila 2006). One way of aligning oneself with whiteness is through formal education and speaking proper Portuguese, which one could achieve by investing in expensive, private school education and avoiding slang (*giria*) in one's speech (Roth-Gordon 2017). Education was flagged as an important strategy in the past (De Azevedo 1996 [1953]) and remains important—indeed, it is integral to many research participants in terms of how they position themselves as respectable, middle-class people and define themselves as Brazilian citizens. The general assumption is that children should attend good schools in order to get into a good university and get a good job, thereby becoming productive members contributing to their families and to society. To not follow this path is to be the "black sheep" in the family—so there is a moral element to this discourse. I often heard research participants describe family members who, in their eyes, did not take their schooling seriously. Thus, to some degree educational strategies are racialized—persons could "whiten" by remaining dedicated to educational success, while they "blacken" by rejecting this path, though this is not a straightforward process.

For research participants, "education" did not only refer to formal education and language but also referred to cultural ideas about proper socialization, which entailed demonstrably good manners, respect for others (*respeito*), being able to conduct oneself properly in public, and having cultural knowledge. Indeed, informants would often assert, *"Sem educação, não tem nada"* ("without education, you have nothing"). Formal education, "proper" speech and etiquette, cultural knowledge and respectful behaviour towards others allow some access to whiteness and could protect research participants, to some extent, from racism. Body language and being comfortable in white spaces and with white practices (such as travel) all amount to ways Afro-Brazilians could invest in personal whiteness, which allows some access to resources and privileges to occupy certain exclusive spaces. Upwardly mobile black and mixed-race Brazilians are said to be treated by others as whiter and may marry lighter-skinned people, thereby producing lighter-skinned offspring.

In her article about race and class in Salvador, the anthropologist Cecilia McCallum gives an account of a conversation she had with a Bahian friend, a white university lecturer, about whether there was an identifiable black middle class in Salvador. At once, the man said that to him, middle-class people are *branco* (white) even if individuals may be described as *negro*, *preto*, or *moreno* (black, brown, or tan). He never thought of his colleague as *negro*—for him, *negro* referred to people who lived in poor areas of the city (McCallum 2005, 109). While I agree that class—in the sense of occupation, lifestyle, comportment, residential location, and so on—does influence race/colour ascription so that upwardly mobile African descendants are seeing as "whitening" middle-class people, I do not agree that their colour is of no consequence. First, while upwardly mobile Afro-Brazilians can, in some senses, be considered "structurally white" in the eyes of others, they do encounter barriers in some spheres of life—such as racism in the workplace and in intimate relationships (see Chap. 2), and especially in terms of aesthetic values related to physical appearance (see Chap. 7). Second, behavioural practices are not unmoored from their biological grounding, as Brazilian social scientist Patricia Pinho has noted. The fact that blackness and racial mixture is long associated with being unhealthy, lazy, and having a predisposition to criminal behaviour means that these negative characteristics are understood as "being as much based on nature as based on nurture" (De Santana Pinho 2009, 43).

Discourses of Brownness

Brazil's national discourse has, for over a century, been grounded in myths of racial harmony and racial democracy, as has been discussed previously. There is a constant and all-pervasive fabrication and diffusion of nationalist ideology through the media, as television and newspaper companies are often owned and controlled by political elites and powerful politicians who wield influence over major channels of communication. This media exerts strong influence over popular events such as international football competitions, carnival, and Formula One car-racing, transmitting images of a unified Brazilian nation to its audiences. Furthermore, most school children are taught about the historical mingling of the three races (Amerindian, African, and European) in the history of the formation of Brazil. However, it is important to note that such discourses are tempered to some extent in a place like Salvador—the "Black Rome" of Brazil.

Miscegenation became the positive contribution of Brazil to the world, producing a new civilization in the tropics, supposedly resulting in its "racial exceptionalism" (Hanchard 1994). This process of *morenidade* (racial mixing) was believed to dissolve conflicts and overcome differences, creating a Brazilian meta-race. At the same time, the distinctiveness of each kind of *mestiço* and his or her alleged qualities was emphasized by scholars such as Freyre.[14] Freyre encouraged the use of the term *moreno* in Brazil as it came to be associated with a wide range of skin colours. Patricia de Santana Pinho has explored Freyre's use of the term which referenced the Moorish influence on the Portuguese, which was thought of as fundamental to the more "flexible" and "open" character of the Portuguese towards racial and cultural mixture (De Santana Pinho 2009, 46). In Brazil, *moreno* stands for a great range of skin colours, from dark to light, while the extreme ends of the pole (black and white) are rejected. In particular, the icon of the sensual *morena* (mixed-race woman) came to rep-

[14] See Patricia De Santana Pinho, 2006. "Gilberto Freyre e a Baianidade" in Malcolm McNee and Joshua Lund, eds. Gilberto Freyre e Os Estudos Latinoamericanos.

Table 3.1 Racial composition in Brazil, 1880–2000, showing increasing identification with the *pardo* (brown/mixed-race) category on the national census

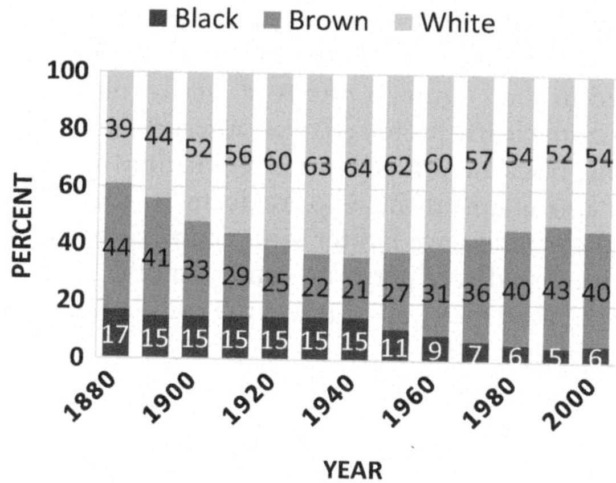

Source: Telles (2004, 30)

resent Brazilian national identity, underscoring the gendered dimension of national discourses of *mestiçagem* (Freyre 1984). Table 3.1 below illustrates an increasing tendency in the early part of the nineteenth century for Brazilians to identify with whiteness, which then begins to decrease in more recent years. It also shows that Brazilians' identification with the *pardo* race) category (brown/mixed race) decreased at first before showing a gradual increase since the 1950s, corresponding to the time period in which the military dictatorship was celebrating the country's mixed-race identity.

National discourses of race mixture are also linked to sexuality, alluding to the erotic myth of the sexual coupling of the different races to form the Brazilian nation. Firstly, Brazil's "birth of the nation" myth revolves around the idea of white male conquistadores who defeated the native warrior, taking possession of his lands, women, and children—later imposing his patriarchal and sexual dominance on black slaves. Thus, non-white women have been brought under white male control, justifying the racial and gendered status quo in Brazil (McCallum 1996, 209). Secondly, valuations of sexual attractiveness are included in how bodies are "read" by others, which tends to result in the commodification of black female bodies, placing them at the bottom of socio-racial hierarchies of desire (Goldstein 1999).

Sociologists have noted a strong identification with being *moreno* over the census category *pardo* in Brazil, as it can encompass a wider range of physical features (Telles 2004; Baran 2007). However, the Brazilian cult of *morenidade* (browning, race mixture) does not mean that all *mestiço* types are valued in the country's myths of racial/cultural mixture. Some types of mixture are clearly preferred over others. In such a racial stratification system, anti-black racism is expressed not only against dark-skinned individuals but also against those who are deemed to have "black" physical traits, even in those who have lighter skin complexion (De Santana Pinho 2009, 40). Thus, mixed-race persons occupy a mobile, flexible, intermediary position, which can move up or down the racial stratification of Brazil's pigmentocracy (Telles 2004).

Discourses of Blackness

A third racial discourse that I will discuss is blackness, which increasingly challenged dominant national discourses. However, blackness, like whiteness, is neither a straightforward natural category nor a straightforward historical category. Rather, it is a complex set of cultural constructions spanning time and space and should be understood in relation to other discourses of race. As African American Studies scholar Kwame Dixon writes, "Afro-Brazilian black social identities are not fixed and static but fluid and dynamic." Furthermore, "it is useful to think about multiple and overlapping or intersectional identities with the history of slavery, racism, racialization, racial violence, social class, gender, religion, skin color, oppression, and memory of Africa as some of the core elements structuring Afro-Brazilian identities." Finally, he states, "within this matrix there are simultaneously competing and contradictory representations of what it means to be black in Brazil and more specifically Bahia by both blacks and non-blacks, Brazilians and non-Brazilians" (Dixon 2016, 23). With this complexity in mind, it is useful to reflect on how these representations of blackness affect the Brazilian Black Movement's efforts to mobilize around a unitary black identity in Brazil.

Black Mobilization and Anti-racist Organizing in Brazil

Since the democratic opening in Brazil in the 1980s, black mobilization and anti-racist organizing have expanded in Brazil, and a dynamic black "civil society" has emerged (Dixon 2016). Before that (from around the 1930s until the middle of the 1970s) anti-racist discourse and large-scale mobilization around issues of racial justice were severely restricted, first under the Estado Novo and later during the military dictatorship from 1964 to 1985. Therefore, Afro-Brazilian activists had to frame their activities in indirect and ambiguous ways, which often took the form of cultural practice (Hanchard 1994, 102). By the mid-1970s, there was stronger opposition to military rule, and by the end of the decade, the country was on a path of liberalization that opened space for the rise of a large cross-section of diverse civil society groups whose members endeavoured to connect their struggles for visibility in Brazilian society to that of social justice and democratic citizenship (Andrews 1996). For example, the São Paulo-based Movimento Negro Unificado (Unified Black Movement) was formed in 1978, arising from earlier forms of activism, such as the Experimental Black Theatre formed in 1945, which had developed a discussion group on racial inequality (Sansone 2003). As Hanchard (1994) has discussed, black activism in Brazil involved a series of movements with distinct ideological commitments and political strategies. In Salvador, these groups included the *blocos afros* (carnival groups) Ilê Aiyê and Olodum, founded in 1974 and 1979 respectively; the Movimento Negro Unificado (MNU), which set up a cell in Salvador; diverse groups of university students; black trade union and political leaders; religious leaders; and those involved in dance, art and theatre. They cultivated new forms of discourse and expanded the struggle for citizenship rights, engaging in grassroots mobilization and enlivening debates about race, gender, and inequality—while at the same time, validating Afro-Brazilian cultural and aesthetic formations.

By 1988, Brazil was officially a democracy with a new constitution; the year also marked 100 years since the formal abolition of slavery. The timing was right for Black and Women's Movements—as well as other civil society groups—to position themselves as serious political actors, articulating long-standing issues of racialized and gendered discrimination and inequalities in the country (Caldwell 2007; Dixon 2016). Some examples include the Steve Biko Cultural Institute (later changing to The Steve Biko Cooperative), formed in Salvador in 1992. Setting up their headquarters in Pelourinho, they aimed to provide educational opportunities to socially excluded Brazilian youth, as well as to prepare them for Brazil's notoriously difficult university entrance examinations, known as the *vestibular*. The founders have a long-standing concern with the radical transformation of education, with a desire to cultivate a critical consciousness and racial pride in the students attending the programme. This includes engaging students in debates about black identity, African history, Afro-Brazilian culture, education, and affirmative action. In 1993, Didá

Educational and Cultural Association was created by Neguinho do Samba, a former *mestre* (master) of Olodum. A non-profit organization, its mission is to educate and empower women and girls through the arts—with a special focus on poor, black women. It also functions as a carnival group and band, while providing free educational activities and training opportunities.

In terms of religious organizations, the Centro Arquidiocensano de Articulação da Pastoral Afro (CAAPA) was established in Salvador in 1999 (Selka 2007, 34). Grounded in the ideas of liberation theology, CAAPA sought to address racism and underrepresentation of blacks within the Catholic Church, to raise black consciousness and reaffirm Afro-Brazilian identity, and to dialogue with the Candomblé community. Protestant and Evangelical groups also began engaging in the new politics of race. For example, Burdick (2005) noted that groups devoted to issues of race and racism began emerging between the 1980s and the 1990s, mainly in Rio de Janeiro and São Paulo, but soon spreading to other states, while becoming more organized and linking themselves with the Black Movement. These examples of black mobilization were important in terms of challenging racism in Brazilian society, ushering in a new social-political landscape, characterized by a more radical discourse of race, black consciousness, Afro-referenced identities, social movement activism, and anti-racist politics.

Outside of Brazil, the independence movements in the Portuguese colonies (especially in Africa) as well as the North American civil rights movement and Pan-Africanist ideology had an impact on Brazil. According to Sansone (2003), dialogue and interchange across the African diaspora and Pan-Africanist discourses mixed with the Brazilian local reality, performing an amalgamating work and giving rise to different forms of identity politics and ways of being black. Brazilian black activists fought against white hegemony and racial discrimination, its most radical section promoting ideas of political and cultural separation. However, the focus from very early on was on Afro-Brazilian cultural forms, perhaps the result of the leadership being drawn from samba schools, associational and religious groups, and artistic arenas (Hanchard 1994, 99). As Candomblé increasingly gained prestige and power in Salvador's public arena, black activists began to promote Candomblé as representative of Afro-Brazilian cultural heritage and resistance, especially after the anti-syncretic movement within the religion itself. Therefore, Candomblé could be deployed as a symbol of Brazilian creativity and racial/cultural fusion (discourses of hybridity) as well as being symbolic of a more essentialized identity (discourses of ethnic absolutism/purity).

In recent years, the Black Movement has transformed itself into several smaller entities, employing a non-governmental organization model with a small core of full-time, paid activists whereas before, it was run mainly by volunteers. These NGOs can attract funding and direct their energies towards specific goals. Therefore, we now have professional activists, designing policies to bring to government. This organizational change from general awareness raising and promoting pride in black identity to public policy was a strategic decision made. Perhaps it was easier to convince public servants and the government, a smaller audience, versus trying to gain public approval and support and change attitudes and behaviours. Indeed, the Black Movement, because it comprises a section of the black community that has ascended socially, tends to be associated with an intellectual perspective that has not attracted the broader masses. It appears that the movement's polarized vision of the Bahian racial situation (divided rigidly between blacks and whites), along with its position of radical political militancy, strikes a discordant note with many peoples' everyday experiences, practices, and customs (Bacelar 1999, 97).

Meanwhile, international pressure has increased for the adoption of human rights and multicultural policies into Latin American political systems. It has now been roughly 60 years since the Declaration of Human Rights was made by the United Nations and this has been followed by legal reforms in most countries. Within this context, it has been possible to "package" black rights as a human rights issue, lending legitimacy and currency to debates about racial inequality in Brazil (Burdick 2005). Therefore, the Black Movement has been able to draw on wider discourses about citizenship and the way that democracy should be inclusive of black and indigenous groups, what has been termed "multicultural citizenship" (Kymlicka 1995). In both public and private spheres, the result of decades of mobilization has seen the coming of age of the Brazilian Black Movement in political terms, especially with the entry of Afro-Brazilians into decision-making circles which is a monumental achievement. Over the past decade, for example, blacks have made slow inroads into politics and have been elected to local, state, and federal office—although their overall representation at each of these levels is still very low.

The opening up of spaces within the government addressing issues of race and racial inequality has also contributed to recent forms of racial consciousness. Since the mid-1990s, the idea that public policy should seek to address issues of racial inequality in the Brazilian society has been increasingly voiced by government officials. By the early 2000s, the Brazilian government committed to the implementation of affirmative action policies, especially in Brazilian public universities—long regarded as institutions which favoured privileged students (Htun 2004; Burdick 2005). The well-known black politician, Benedita da Silva, helped to pave the way for the implementation of affirmative action in state universities when she was Governor of Rio de Janeiro. During this time, the Federal University of Bahia (UFBA) announced that 43 per cent of vacancies would be reserved for students from the public education system and black students (Dos Santos 2010, 760). Once in office, Lula also committed his government to fighting racial inequality, naming three prominent Afro-Brazilians to his cabinet: Gilberto Gil, Minister of Culture; Marina Silva, Minister of Environment; Matilde Ribeiro, Special Secretary for the Promotion of Racial Equality (Dos Santos and Anja 2006, 39). In addition, a Municipal Secretary of Reparations (Secretaria Municipal de Reparação) was established in Salvador.

Despite research that has pointed to a lack of broad support among the population for Black Movement discourse and activities (e.g. Sansone 2003; Twine 1998), the discourse of the movement (especially as it related to black pride and addressing issues of racial discrimination and racial justice) has filtered down through the population so that a younger generation increasingly identifies as being black—or they are either challenged to think of themselves in that light or may accept or reject these discourses to varying degrees (see Baran 2007).[15]

Racial Consciousness and Black Identities

Different forms of racial consciousness emerged from the real, lived histories of residents of the city. Liberdade, a community that was created around the first half of the twentieth century, has been integral to the construction of recent discourses of blackness. Liberdade is understood to be a place with a special history of black organization and resistance. As

[15] Aspects of these arguments were developed in earlier publication by the author. Please see: "A Beleza abre Portas: Beauty and the Racialized Body among Black Middle-Class Women in Salvador, Brazil," (July 17, 2013, Feminist Theory 14: 203–218) and "Negotiating inequality: The Black Middle Classes in Salvador, Brazil," in People, Money and Power in the Economic Crisis: Perspectives from the Global South, 106–128, edited by Keith Hart and John Sharp (2014, Berghahn Press).

Magalhães (2006) has argued, the insertion of several of Liberdade's inhabitants into the formal sector of Salvador's economy seems to have been crucial to the development of racial solidarity and resistance as well as residents' engagement with Bahian political movements of various kinds, such as trade unions. Traditionally, many of Liberdade's inhabitants worked in the old industries connected to the port, commerce, and food industry. More recently, the dynamic industries such as petroleum, chemicals, and metallurgy attracted workers in this area (Agier 1990, 5–6). A considerable portion of Liberdade's population has also worked in the precarious construction industry, commerce, or informal sector. Others have jobs in the city, with women often working as receptionists, office workers, domestic servants, and, more recently, the telecommunications industry.

It was not uncommon for a large group of people to work for the same company, thus Agier (1992) has argued that people began to relate to this labour market, employing a new idiom of "individual choice and strategy...related to racial solidarity" (cited in Magalhães 2006, 52). Thus, "becoming black" seemed to represent a clear, strategic break with the past (Magalhães 2006, 53). What seemed less clear in this analysis, however, is the basis for these new forms of racial solidarity and identity. (In other words, what did this change from, and what caused the change?) A possible answer may be found in a later article written by Michel Agier, in which he argued that contrary to their hopes of being integrated into a modern, dynamic industrial economy with the possibility of improving their lives, most of these new industrial workers often encountered racism in their work environments and a cut-off point beyond which they could not progress (Agier 1994, 7–8). He identified several different responses to this situation in his informants' trajectories—it was typical that a person may employ a combination of them.

Firstly, a person may increasingly identify with the cultural Afro-Brazilian universe. Often, this had a deeper weight when someone had a family member involved in Candomblé or carnival groups. This would lead to a more culturalist discourse on blackness. Secondly, a person might adopt a more radical and political version of the first response. The emerging group of black workers in Liberdade apparently had more money and time to spend organizing cultural and leisure activities, such as carnival *blocos* and *afoxés* (such as *afoxé* Filhos de Gandhy, which was formed by stevedores, dock workers, and other working-class people from the neighbourhood) as well as several contemporary black militant groups. Powerful new forms of black culture were created which echoed the struggle for civil rights in the United States and the Independence movements in the Portuguese colonies in Africa. The professionalization and commercialization of certain aspects of Afro-Brazilian culture (such as Capoeira and music) and the emergence of cultural organizations such as Ilê Aiyê opened up new opportunities for social mobility and options for positive social identities for blacks, yet sometimes in limited and frustrating ways. Thirdly, a person may appear to do exactly the opposite: to try avoid interpersonal situations that could involve racial difference; to avoid talking about race or racism; to distance themselves from the radical discourse of the Black Movement; or to deny personal experiences with racism.

Investment in the working identity was one way of distancing oneself from situations that could be potentially disrespecting. It also tends to lead to their symbolic association with whiteness and the social and moral values attributed to whiteness. The silence of these workers on race and the Black Movement may derive from the notion of blackness being associated with negative and traditionally low prestige activities. Thus, Agier (1994, 10) argues that those involved in the fight for a new status tended to stress the importance of respect, being known for their professional competency or capacity to bring up a family and to reject the social significance of colour. These workers were not satisfied with the racial and social identity ascribed to them by dominant, hegemonic discourses. Therefore, they tended to identify with work or the company, preferring to be acknowledged for their personal qualities, such as their sincerity, dedication, or popularity. The workers with this kind of trajectory tended to achieve in union positions.

Some residents became involved in the community life of Liberdade and developed powerful forms of racial consciousness, with Candomblé as its central axis. Cultural practices became ideology and politics in the construction of these discourses, and they had a powerful appeal because of their proximity to peoples' everyday lives. Several prominent personalities were important here. The well-known Mãe Hilda—a *mãe de santo* (spiritual mother)—supported the creation of Ilê Aiyê by her son and a number of his friends and family members. The idea behind Ilê Aiyê was to provide a *bloco* only for blacks. In a dynamic form, a sense of African roots was revived, and black pride was promoted in the form of the practices, hairstyles, clothes, and the drum music of Ilê Aiyê. When this *bloco* started to go out in carnival in 1975, they attracted a lot of attention and some people were suspicious, labelling the group as racist. It should be noted that during this period, Brazil was under military rule and they were against the incitement of racial antagonism, particularly any kind of organized, political entity. However, there was no attempt to dismantle Ilê Aiyê. Now their *bloco*, with beautiful women dancing on top of high-powered trucks surrounded by African-clothed, dark-skinned Bahians (as well as many African Americans), are an exciting feature of carnival as they emerge from Liberdade into Campo Grande (a park in the City Centre) in the early morning hours singing their powerful anthem song:

> *What is the bloco?/ I wish I knew/ It's the Black world/ Parading here for you/ We are the dark Creoles/ Happy and proud/ Of our Kinky hair/* We are the Black Power/ White, if you knew/ The worth of the Black man/ You'd bathe in tar/ And become Black like us.[16]

Since the 1970s, Ilê Aiyê has evolved into a well-funded, non-governmental organization in the Curuzu neighbourhood linked to the Black movement, offering classes to underprivileged children and promoting African culture, heritage, and Black pride. Ilê Aiyê is well known for its annual "Beleza Negra" competition where a dark-skinned female with the most charismatic on-stage personality and ability to dance Candomblé rhythms is chosen. She is then crowned as Queen of their carnival parade. Ilê Aiyê's *ensaios* (carnival rehearsals) are also extremely popular events. The work of this organization has been important to the construction and dissemination of a positively valued black identity in Salvador. Figures 3.2 and 3.3 are photographs taken by the author of Ilê Aiyê headquarters in Liberdade.

Race and Genetics

The persistence and even re-emergence of biological versions of race or racialized constructs are particularly evident in recent genetics research in Brazil and ensuing public debates. The recent appearance of DNA ancestry testing services has the potential to lead to increasingly deterministic ways of imagining ethnic and racial heritage. Sheriff (2001) has already argued that there exists in Brazil a conception of race that is more or less genetic, based on a notion of descent. However, the development of genetics as a field of science in Brazil might also be contributing to new angles on the racial question. At the same time, it is important to note that

[16] Cited in Magalhães (2006, 54).

Fig. 3.2 Ilê Aiyê, a well-known Afro-bloco, located in Curuzu, Liberdade, Salvador. (Source: Doreen Gordon)

DNA ancestry testing is a field of contested knowledge, where the limits of scientific knowledge are debated (Bolnick 2008; Bolnick et al. 2007).

Genetic labs tend to concentrate on paternity testing, but some have been mapping the genomes or genetic composition of local populations. It is believed that Brazil offers a unique opportunity to study the effects of "genetic admixture" which is best carried out in populations that have intermixed for centuries, such as Latin American populations (Pena et al. 2000; Sans 2000). In some countries, genetic mapping of different "populations" is felt to be important for combating diseases thought to have a

Fig. 3.3 The interior of Ilê Aiyê where carnival rehearsals are held, Liberdade, Salvador. (Source: Doreen Gordon)

genetic component—an idea which has been around for some time in relation to certain illnesses such as sickle cell anaemia, diabetes, and syphilis and their supposed link to black populations (Calvo-Gonzalez and Rocha 2010; Fry 2005; Laguardia 2005; Diniz and Guedes 2006). In this sense, slippage between interpreting populations as social and biological entities can occur, potentially underwriting social categories with biological explanations (Kent and Wade 2015).

Within Brazil, the influences of this discourse on genetics can be seen in recent scientific work. A project co-ordinated by a team of geneticists at the Universidade Federal de Minas Gerais in Belo Horizonte aimed to shed light on the genetic origins of Brazilians based on the sequencing of parts of mitochondrial DNA and the Y chromosome. The researchers aimed to identify what percentage of genes were Amerindian, European, or African in 200 men from different regions of Brazil who classified themselves as white (Pena et al. 2000, 20–21). The research showed that at least half the African descendants tested had European ancestry on the father's side. Meanwhile, the mother's side revealed a predominantly African and Amerindian genetic makeup, appearing to confirm Brazil's mythical origins in the union between a white colonizer and an Amerindian or African woman. The results of the project were widely disseminated in the media, as part of the Afro-Brazilian Roots Project, commissioned by BBC Brasil, focusing in particular on the genetic profiles of nine black celebrities such as the gymnast Diane dos Santos and Neguinho da Beija-Flor of a famous Rio de Janeiro samba school.[17] Neguinho da Beija-Flor (whose name roughly translates into "little black" or "blackie") is a shining symbol of Brazil's black community and culture, where he plays a leading role in the Beija-Flor samba school. His genetic profile sparked debates on the racial question in Brazil, as his test indicated that he was 67.1 per cent European ancestry and only 31 per cent African. Pena considered this a typical case of the high level of genetic mixture in the population that his studies have revealed for some time (Pena et al. 2000). Neguinho, who is very dark-skinned, responded, "People will think I'm joking if I tell them this." His case sparked heated debates about affirmative action policies aimed at Afro-Brazilians, such as university quotas and differential health policies. Certainly there were some segments of Brazilian society—including centre-right and far-left political parties, the mass media, and some social scientists—who considered race-based affirmative action policies inappropriate for a country as racially mixed as Brazil, preferring class-based criteria for inclusion instead (Fry et al. 2007).

While the articles and videos featured on the "Afro-Brazilian Roots" website appeared to undermine the idea of race itself, Brazil's Black Movement dismissed the relevance of genetic testing by arguing, first, that in Brazil, race functions as a social rather than biological category, and second, that racial classification and discrimination in the country is largely based on physical appearance, rather than genotype (Kent and Wade 2015).

Meanwhile, other members of the public have been interpreting genetic studies in varied ways, including in ways that underscore a more essentialized identity. For example, sickle cell anaemia has received greater attention in recent decades in Brazil, especially because of the Black Movement and the development of a political camp of action within the Ministry of Health called "Saude de População Negra" (Health of the Black Population). In this context, persons can employ genetic proof of black

[17] See Silvia Salek, "BBC delves into Brazilians' Roots," BBC Brasil, July 10, 2007, http://news.bbc.co.uk/2/hi/6284806.stm. Also see the original website which is now archived on the BBC, https://www.bbc.com/portuguese/noticias/cluster/2007/05/070427_raizesafrobrasileiras

identity via being diagnosed with an illness such as sickle cell anaemia. This scenario may indicate that biology is becoming the battle ground on which questions of race and identity are fought out and transformed (Skinner 2006; Wade 2007). Thus, the tracing of identities through genetic material could potentially become a political act, reflecting a new kind of biosociality (Rabinow 1992, 241), pointing to notions of biological citizenship.

Racial Discourse in Everyday Life

A key ethnography in English that placed upwardly mobile Afro-Brazilians and their understandings and experiences of race at the centre of its analysis was *Racism in a Racial Democracy* (1997), written by France Winddance Twine. Based on more than 50 life histories with Afro- and Euro-Brazilians, her book analysed how upwardly mobile Afro-Brazilians living in a rural town in south-eastern Brazil conceptualized and responded to racial disparities—usually in ways that reinforced the dominant ideology of "racial democracy" in Brazil. In this section, I push the analysis further, taking into account the effect on Afro-Brazilians of 30 years of black mobilization, a deepening critique of ideas of "racial democracy" and changing discourses about race in Brazil.

Similar to some researchers writing on race relations in Brazil (Twine 1998; Sheriff 2001), my ethnographic material supports the presence of competing—but at times complementary—racial discourses that emerge in particular instances. Indeed, racial discourse is embedded in everyday interactions, but are often codified or masked. Thus, in some social interactions, whiteness or blackness may be highlighted as visible discourses. However, at other times, distinctive race/class projects surrounding mixed-race brown identities emerged. Nevertheless, these discourses are not uncontested in Bahia. For example, some residents question national ideologies and discourses of race mixture in Brazil and are critical of dominant political and class interests in the state of Bahia.

The symbolic association between whitening and social mobility in relation to upwardly mobile Afro-Brazilians is also not a direct or simple one, nor does it appear to have been in the past (Pierson 1942; De Azevedo 1996 [1953]). Research participants sometimes argued that what was perceived as distancing oneself from blackness was not the result of a conscious motive to do so—such as moving to another part of the city relatively isolated from former ethnic networks or the desire to break away from responsibilities to one's less well-off kin. One informant, a lawyer coming from a humble background who now lives in one of the modern areas of the city, stated that he prefers to distance himself from his "crazy" relatives because they do not understand the nature of his career and family commitments. The question of choosing a white or lighter-skinned partner is one of the more eye-catching strategies of black social mobility and is a sensitive topic for discussion. I will discuss this further in Chap. 6, when I explore the question of interracial marriage.

Acceptance of blacks into higher social positions often involves conditional acceptance which could become more intolerant of overt displays of blackness, as one learns to "play the game," as Jorge Carvalho put it. As one successful black businessman—the manager of an elite chain of luxury grocery stores—said to me, he would never do anything obvious such as go to a party of mostly white friends dressed up in African garb. "Playing the game" or following the social script in similar settings could involve the full or partial acceptance of the inferiority of blackness so that an upwardly mobile person may adopt disparaging attitudes towards other blacks or to blackness in specific situations, such as when *pessoas de cor* (people of colour) in Azevedo's 1950s study looked down on uneducated black people. Thus, blackness remains a primary symbol of lower-class status which often must be specifically re-negotiated in order to avoid negative stigma.

The national ideology of racial democracy and racial mixing could also be traced in the discourses of my research participants, especially among the older generation, through utterances such as *"somos todos mestiços"* ("we are all mixed") or *"somos todos Brasileiros"* ("we are all Brazilian")—even as they also identified as black Brazilians and acknowledged the existence of racial discrimination. In discussing the racial attitudes of her parents' generation, Helena Teixeira, who was in her mid-60s, explained that older family members did not see themselves as black. For the Teixeira family, having indigenous and white ancestors appeared to refine their social status. Indeed, in Salvador's socio-racial hierarchy, different combinations of physical features were valued differently—especially hair texture which functioned as a primary signifier of the degree of closeness to whiteness or blackness.

Nevertheless, the ideas of the Black Movement have a certain sway in Salvador where increasingly, lighter-skinned Brazilians may identify as black, wearing their hair in styles that would be referred to as *cabelo black* ("natural" hairstyles such as braids, dreadlocks, Afros), wearing Hip Hop fashion associated with an American (now becoming Brazilian) style of black consumption visible in the mass media and on foreign music videos. It is important to recognize a shift here—whereas historically, the higher strata were the importers of foreign culture, today a powerful mass media conveys these messages to a public with wider access to televisions, radios, and the internet. Thus, foreign influences have mixed with local ones, influencing the subaltern categories among blacks. Yet at the same time, the mobility of blackness and its current inscription as *legal* ("cool") are not resources available equally to all. For example, one informant challenged the black identification of a cousin of hers, who she felt was *falsa* ("fake") because she only celebrated her well-rounded *bunda* (bottom) as being black.

It may be said that traditional practices of distancing oneself from blackness have decreased. This is not just happening at the abstract level of policy changes on racial issues but also in informant's everyday lives at the level of personal relationships and identity, evidenced by the increased numbers of people identifying as *preto* on the national census (Nobles 2000; Petruccelli 2007). This may explain the tendency of research participants to identify as black and to contest other peoples' racial ascription of themselves as lighter-skinned. For example, one informant—Bia Sanchez—described herself as *negra*, stating that she rejects when others refer to her as *marron bombom* ("brown sweet") and *morena* (brown, mixed race)—an ascription that she feels they make because of her long, straight hair. Her family and friends may have preferred to describe her in what they perceived as racially favourable terms, because it expressed their future desires for her social mobility, given her education and her marriage to a light-skinned Bahian man.

On the other side of the city, in the well-to-do seaside neighbourhood of Barra, I also came across research participants that strongly identified with blackness. I interviewed three sisters (Sandra, Anita, and Anabela Teixeira), two of whom preferred not to chemically process or straighten their hair, a practice that is expected of upwardly mobile Afro-Brazilians to achieve a "whiter" appearance. The sisters emphasized that they participated in Afro-Bahian culture—for example, by frequently visiting Pelourinho where Afro-Bahian culture is celebrated or going to "Beleza Negra" (Black Beauty) contests and Ilê Aiyê rehearsals in the working-class district of Liberdade. At the same time, the sisters described their family heritage as mixed (meaning that their mother's side of the family is darker-skinned and was mixed with indigenous heritage, while their father is a descendant of Portuguese migrants) yet even his family is mixed. The mother of the three sisters explained that her daughters were born with different combinations of skin colour, hair texture, and eye colour, representing the different racial combinations one can find in Brazil. Thus, family members valued their blackness within long-standing national discourses of racial mixture.

Other research participants described their black identity as a process of coming into realization that they were black through some personal experience. For example, one mother who had thought of herself as *morena* for years increasingly defined herself as *negra* in the face of the discrimination her very dark-skinned daughter faced in the predominantly white schools and neighbourhoods in which she grew up.

As stated previously, downwardly mobile whites or those living among blacks may symbolically darken in consequence, referring to themselves as *mestiço* to highlight their solidarity with the black community. Túlio Santana's wife discovered that her family was mixed race and therefore reclassified herself from *branca* (white) to *morena clara* (light brown). As Túlio and their children are darker-skinned, they identify as *negro* (black)—so it did not appear to be the case that they distanced themselves from a black identity or that they reclassified themselves as lighter because of their mother's skin colour.

Another research participant in his 30s, who calls himself DJ Funk Soul, styled himself as a deejay of *Música Black* (Black Music). He appealed to a more Pan-African, diasporic black identity that at the same time is not exclusionary. Therefore, DJ Funk Soul samples from other Brazilian and international musical styles such as techno and electronic music which more often put him in contact with white audiences. Indeed, a group of Brazilians visiting from Rio de Janeiro once expressed their disappointment at attending one of DJ Soul Funk's parties. Known in their circles for his distinctive *Música Black* selections, they were disappointed to hear him play mostly techno and electronic music. They complained bitterly and described him as being one of those sell-out black people, or a traitor to his race. Clearly, DJ Soul Funk was not performing blackness as was expected of him. Hence black dispositions are culturally essentialized and made authentic—or inauthentic—depending on an individual's performance of race.

Other interlocutors described their racial identity as being based on genetics. For example, Mariela Silveira, a public sector worker in the Ministry of Health, described a process of coming into a realization that she was black through her daughter's illness (her daughter has sickle cell anaemia). Mariela Silveira would be described by most Soterpolitanos (Salvador's residents) as *morena*, but she insists that she has always identified with black people, even though she acknowledges that they did not always accept her as a brown-skinned woman with curly hair, especially within the Black Movement circle. She recalled one incident when she wanted to march with Ilê Aiyê, but they explained to her that she was not dark enough. When one of her daughters was born with sickle cell anaemia, she explained that she did not know much about it. As she educated herself on the illness, she came to an awareness that the illness particularly affects black populations. At this point, she said, her increased identification with being black coincided with her practical work life and day-to-day living, taking care of her daughter. She organized, along with a group of other government health workers, a campaign on issues of racial inequality and access to health care in Salvador. The campaign was initiated at the right time, in an atmosphere of changing public attitudes and government policies to race that paved the way for the success of the working group. She is now firmly embedded within the networks of the Black Movement where she got a chance to say to the President of Ilê Aiyê, "I did a blood test and you will see that I am black!" Here, Mariela Silveira felt she had finally triumphed by the genetic evidence of having a recessive gene for sickle cell anaemia in her body. This example constitutes a particular process of racialization that is reminiscent of earlier discourses of race in Bahia that symbolized ethnic absolutism (e.g. Nina Rodrigues' writings that located racial difference firmly within the body, in hereditary ways).

So far, the examples above suggest that individuals or groups can freely choose their racial identity or employ different racial discourses in varied contexts. On the contrary, there are situations in which race is inscribed and sticks unto certain bodies, in such a way that it is difficult to perform other identities. This theme is explored in a brilliant, long-standing musical play by the Olodum Theatrical Band called "Cabaré de RRRRRaca" (Cabaret of Race). During my fieldwork in Salvador, I went with a few friends to see this play, directed by Márcio Meirelles and Chica Carelli, which was showing at the Vila Velha Theatre in the city centre. The play explored the question of race in a humorous way that also communicated serious social commentary on racial stereotypes, racism, and black aesthetics. The play raised issues of how the black person in Salvador is both a consumer and is also consumed as an object. For example, the black man is stereotyped as *super negão*—a strong, muscular black man who is macho, sexually virile, and desirable. The myth of the strong black man is symbolically linked to physically arduous, manual labour—rather than non-manual or intellectual labour. His desirability plays on racialized notions about black men being more physically and sexually interesting than white men. Indeed, during the play, black male actors come out on the stage naked, displaying their oiled bodies and penises to the audience. Thus, the play explored how some racialized discourses stick to certain bodies, so that it becomes almost impossible to escape other peoples' categorizations or to perform other kinds of identities (such as black men as intellectual professionals). This was an issue that Rodrigo Figueiredo, a journalist, discussed with me many times, in informal conversations we had while sipping cups of coffee in his kitchen or having drinks at the local bar. He was concerned with countering these stereotypes of the black male both in his personal life and at work. For example, he deliberately dressed in unconventional ways (such as wearing pastel, Cuban-style shirts, wooden necklaces, or bracelets) and cultivated intellectual and artistic friends.

Discussion

In this chapter, I mapped out key discourses of race which are constantly being re-worked and transformed in the dynamic, unequal context of Salvador, at times pulling on ideas from wider national and transnational contexts. These models of race also draw on local discourses about hybridity (or *mestiçagem*) and purity (ethnic absolutism). The point was to show different strands or varieties of these discourses which are located along a continuum, influenced by historical processes and transformations as well as the interaction of particular class and race "projects" articulated by a wide variety of social actors, including the state, elites, intellectuals, activists, and ordinary people themselves.

Following Michael Omi and Howard Winant (2014 [1986], 1994), I explored racial discourses at different levels, revealing how racial meanings are constantly being transformed by political struggle and that people can engage with and contest the definition of race both at the macro and micro level. Racial formation theory makes room for the complex and contingent ways that research participants understood, performed, and subverted racial identities, mobilizing knowledges they gleaned from their particular, local situation as well as a range of wider ideas and practices circulating in their public spheres. However, it is also important to think about the ways in which black people are interpellated in the socio-racial hierarchy of Salvador such that they may not have a choice in how they are positioned, valued, or judged by others.

References

Agier, M. 1990. Espaço Urbano, Família, e Status Social: O Novo Operariado Baiano nos seus Bairros. *Cadernos CRH* 13: 39–62. Salvador: Universidade Federal da Bahia, Centro de Recursos Humanos.

———. 1992. Novos Status e Outros Novos Negros: Questões de Identidade entre Trabalhadores Baianos. *Série Toques* (1). Salvador: Universidade Federal da Bahia, Centro de Recursos Humanos.

———. 1994. 'Classe' ou 'Raca?' Socialização, Trabalho e Identidades Opcionais. *Bahia: Análise e Dados, O Negro* 3 (4): Marcha.

Andrews, R. 1996. Brazilian Racial Democracy, 1900–90: An American Counterpoint. *Journal of Contemporary History* 31 (3): 491–525.

Bacelar, J. 1999. Black Brazil. In *Black Brazil: Culture, Identity, and Social Mobilization*, ed. L. Crook and R. Johnson. Los Angeles UCLA Latin American Center Publications, University of California.

Baran, M. 2007. "Girl You Are Not Morena. We Are Negras!": Questioning the Concept of "Race" in Southern Bahia, Brazil. *Ethos* 35 (3): 383–409.

Beckford, G. 1972. *Persistent Poverty: Underdevelopment in Plantation Economies of the Third World*. London: Oxford University Press.

Bolnick, D.A. 2008. Individual Ancestry Interference and the Reification of Race as a Biological Phenomenon. In *Revisiting Race in a Genomic Age*, ed. Sandra Lee, Barbara Koenig, and Sarah S. Richardson. New Brunswick: Rutgers University Press.

Bolnick, D.A., D. Fullwiley, T. Duster, R.S. Cooper, J.H. Fujimura, J. Kahn, J.S. Kaufman, J. Marks, A. Morning, A. Nelson, P. Ossorio, J. Reardon, S.M. Reverby, and K. Tallbear. 2007. The Science and Business of Genetic Testing. *Science* 318 (5849): 399–400.

Borges, D.E. 1992. *The Family in Bahia, Brazil, 1870–1945*. Stanford: Stanford University Press.

Burdick, J. 2005. Why Is the Black Evangelical Movement Growing in Brazil? *Journal of Latin American Studies* 37: 311–332.

Butler, Kim D. 1998. Freedoms Given Freedoms Won: Afro-Brazilians in Post Abolition Sao Paulo and Salvador. New Brunswick, New Jersey and London: Rutgers University Press.

Caldwell, K.L. 2007. *Negras in Brazil: Re-envisioning Black Women, Citizenship, and the Politics of Identity*. New Brunswick: Rutgers University Press.

Calvo-Gonzalez, E., and V. Rocha. 2010. "Está no Sangue:" A Articulação de Ideias Sobre 'Raça', Aparência e Ancestralidade Entre Famílias de Portadores de Doença Falciforme em Salvador, Bahia. *Revista de Antropologia* 53 (1): 277–320.

Centro Cultural Banco do Brasil. 2005. *Por Ti América: Arte Pré-Columbiana*.
Collins, J.F. 2004. 'X' Marks the Future of Brazil. In *Off Stage/ On Display: Intimacies and Ethnographies in the Age of Public Culture*, ed. A. Shryock. Stanford: Stanford University Press.
———. 2015. *Revolt of the Saints: Memory and Redemption in the Twilight of Brazilian Racial Democracy*. Durham: Duke University Press.
Dávila, J. 2006. *Diploma de Brancura: Politico Social e Racial no Brasil, 1917–1945*. São Paulo: Editora UNESP.
Davis, D.J. 1999. *Avoiding the Dark: Race and the Forging of National Culture in Modern Brazil*. Aldershot: Ashgate.
De Azevedo, T. 1996 [1953]. *As Elites de Cor numa Cidade Brasileira*. Salvador: Editoria da Universidade Federal da Bahia.
De Holanda, S.B. [1936] 2006. *Raízes do Brasil*. São Paulo: Companhia das Letras.
De Santana Pinho, P. 2006. Gilberto Freyre e a Baianidade. In *Gilberto Freyre e os Estudos Latinoamericanos*, ed. Malcolm McNee and Joshua Lund. Pittsburgh: Instituto Internacional de Literatura Iberoamericana, University of Pittsburgh.
———. 2009. White But Not Quite: Tones and Overtones of Whiteness in Brazil. *Small Axe* 13 (2): 39–56.
———. 2010. *Mama Africa: Reinventing Blackness in Bahia*. Durham: Duke University Press.
Diniz, Deborah and Guedes, Christiano. 2006. Genetic Discrimination: Sickle Cell Trait in Brazil. Physis: Revista de Saude Coletiva 17 (3): 501–520.
Dixon, K. 2016. *Afro-Politics and Civil Society in Salvador da Bahia, Brazil*. Gainesville: University Press of Florida.
Dos Santos, S.A., and O.C. Anya (trans.). 2006. Who is Black in Brazil? A Timely or a False Question in Brazilian Race Relations in The Era of Affirmative Action? Latin American Perspectives 33 (4): 30–48.
Dos Santos, R.E. 2010. New Social Activism: University Entry Courses for Poor and Black Students. In *Brazil's New Racial Politics*, ed. Bernd Reiter and Gladys L. Mitchell, 197–225. Boulder: Lynn Rienner Press.
Faiola, A., and M. Lopez. 2018. How Jair Bolsonaro Entranced Brazil's Minorities—While Also Insulting Them. *Washington Post*, October 24. https://www.washingtonpost.com/world/the_americas/how-jair-bolsonaro-entranced-brazils-minorities%2D%2Dwhile-also-insulting-them/2018/10/23/a44485a4-d3b6-11e8-a4db-184311d27129_story.html. Accessed 6 Aug 2021.
Fanon, F., and R. Philcox (trans.). [1952] 2008. *Black Skin, White Masks*. New York: Grove Press.
Fischer, B. 2004. Quase Pretos de Tão Pobres? Race and Social Discrimination in Rio de Janeiro's Twentieth Century Criminal Courts. *Latin American Research Review* 39 (1): 31–59.
Freyre, G. 1984. Uma Paixão Nacional. *Playboy*, no. 113 (December). The following link is no longer available: http://www.releituras.com/gilbertofreyre_

bunda.asp. However, the article is extracted on the following website, "Conto Basileiro: Contos, Crônicas e Poesias de Autores Brasileiros." https://conto-brasileiro.com.br/bunda-paixao-nacional-artigo-de-gilberto-freyre/. Accessed 6 Aug 2021.

Fry, P. 2005. O Significado da Anemia Falciforme no Contexto da 'Politica Racial' do Governo Brasileiro 1995–2004. *História, Ciências, Saude. Manguinhos* 12 (2): 347–370.

Fry, P., Y. Maggie, and M.C. Maio. 2007. *Divisões Perigosas: Políticas Raciais no Brasil Contemporâneo*. Rio de Janeiro: Civilização Brasileira.

Goldstein, D. 1999. "Interracial" Sex and Racial Democracy in Brazil: Twin Concepts? *American Anthropologist* 101: 563–578.

Hafiz, J. 2016. The Cabula 12: Brazil's Police War against the Black Community. *Aljazeera America*, February 25. http://america.aljazeera.com/watch/shows/america-tonight/articles/2016/2/25/the-cabula-12-brazil-police-war-blacks.html. Accessed 6 Aug 2021.

Hanchard, M.G., ed. 1994. *Orpheus and Power: The Movimento Negro of Rio de Janeiro and São Paulo, Brazil, 1945–1988*. Princeton: Princeton University Press.

Hartigan, J. Jr. (2000). Whiteness in the Field: Introduction to a Speical Issue of *Identities: Global Studies in Culture and Power* 7 (3): 269–279.

Henfrey, C. 1981. The Hungry Imagination: Social Formation, Popular Culture, and Ideology in Bahia. In *The Logic of Poverty: The Case of the Brazilian Northeast*, ed. S. Mitchell. London: Routledge and Kegan Paul Ltd.

Henriques, F. 1968 (1953). *Family and Colour in Jamaica*. London: MacGibbon and Kee Ltd.

Htun, M. 2004. "From 'Racial Democracy' to Affirmative Action: Changing State Policy on Race in Brazil." *Latin American Research Review* 39 (1): 60–89.

Instituto Brasileiro de Geografia e Estatística. 2012. *Censo Demográfico 2010*. Brasilia: Instituto Brasileiro de Geografia e Estatística.

Kent, M., and P. Wade. 2015. Genetics Against Race: Science, Politics and Affirmative Action in Brazil. *Social Studies of Science* 45 (6): 816–838.

Kymlicka, W. 1995. *Multicultural Citizenship: A Liberal Theory of Minority Rights*. New York: Oxford University Press.

Laguardia, Josué. 2005. Raça, Genética e Hipertensão: Nova Genética ou Velha Eugenia? História, Ciências Saúde-Manguinos, Rio de Janeiro 12 (2): 371–393.

Magalhães, M.S.R.A. 2006. Another Bahia: Cosmopolitanism, Violence, and Sovereignty in Salvador, Brazil. PhD Dissertation, University of Cambridge, Cambridge.

Marx, A.W. 1998. *Making Race and Nation: A Comparison of the United States, South Africa, and Brazil*. Cambridge: Cambridge University Press.

Matory, L. 2005. *Black Atlantic Religion: Tradition, Transnationalism and Matriarchy in the Afro-Brazilian Candomblé*. Princeton: Princeton University Press.

McCallum, C. 1996. Resisting Brazil: Perspectives on Local Nationalisms in Salvador de Bahia. Ethnos 61 (3-4): 207–229.

———. 2005. Racialized Bodies, Naturalized Classes: Moving Through the City of Salvador da Bahia. *American Ethnologist* 32 (1): 100–117.

Moreno Figueroa, M.G. 2013. Displaced Looks: The Lived Experience of Beauty and Racism. *Feminist Theory* 14 (2): 137–151.

Nacsimento, A. do. 1989 [1979]. *Brazil, Mixture or Massacre?: Essays in the Genocide of a Black People*. Dover: The Majority Press.

Nobles, M. 2000. *Shades of Citizenship: Race and Census in Modern Politics*. Stanford: Stanford University Press.

Omi, M., and H. Winant. 1994. *Racial Formation in the United States: From the 1960s to the 1990s*. New York: Routledge.

———. 2014 [1986]. *Racial Formation in the United States*. New York: Routledge.

Otovo, O.T. 2016. *Progressive Mothers, Better Babies: Race, Health, and the State in Brazil, 1850–1945*. Austin: University of Texas Press.

Pena, S.D.J., D.R. Carvalho-Silva, J. Alves-Silva, F.R. Santos, and V.F. Prado. 2000. Retrato Molecular do Brasil. *Ciencia Hoje* 159: 16–25.

Perry, Keisha-Khan Y. 2013. Black Women Against the Land Grab: The Fight for Racial Justice in brazil. Minneapolis and London: University of Minnesota Press.

Perry, K.K.Y. 2019. The Resurgent Far Right and the Black Feminist Struggle For Social Democracy in Brazil. *American Anthropologist* 122 (1): 157–162.

Petruccelli, J.L. 2007. *A Cor Denominada: Estudos Sobre A Classificação Étnico-Racial*. Rio de Janeiro: DP and A Editora Ltda.

Phillips, T. 2007. Antonio Carlos Magalhaes. *The Guardian*. https://www.theguardian.com/news/2007/jul/23/guardianobituaries.brazil. Accessed 6 Aug 2021.

Pierson, D. 1942. *Negroes in Brazil: A Study of Race Contact at Bahia*. Chicago: The University of Chicago Press.

Rabinow, P. 1992. Artificiality and Enlightenment: From Sociobiology to Biosociality. In *Incorporations*, ed. J. Crary and S. Kwinter. New York: Zone Publications.

Reiter, B. 2010. "Whiteness as Capital: Constructing Inclusion and Defending Privilege." In Bernd Reiter and Gladys L. Mitchell (eds), Brazil's New Racial Politics. Boulder, Co and London: Lynne Rienner, 19–34.

Roth-Gordon, J. 2017. *Race and the Brazilian Body: Blackness, Whiteness, and Everyday Language in Rio de Janeiro*. California: University of California Press.

Salek, S. 2007. BBC delves into Brazilians' Roots. *BBC Brasil*, July 10. http://news.bbc.co.uk/2/hi/6284806.stm. Also see the original BBC website which is now archived at: https://www.bbc.com/portuguese/noticias/cluster/2007/05/070427_raizesafrobrasileiras. Accessed 6 Aug 2021.

Sans, M. 2000. Admixture Studies in Latin America: From the 20th to the 21st Century. *Human Biology* 72 (1): 155–177.

Sansone, L. 2003. *Blackness Without Ethnicity: Constructing Race in Brazil*. New York: Palgrave Macmillan.
Santos, J.T. dos. 2005. *O Poder da Cultura e a Cultura No Poder: A Disputa Simbólica da Herança Cultural No Brasil*. Salvador: Editora da Universidade Federal da Bahia.
Schwarcz, L.M. 1993. *The Spectacle of the Races: Scientists, Institutions, and the Race Question in Brazil, 1870–1930*. New York: Hill and Wong.
Seigel, M. 2009. *Uneven Encounters: Making Race and Nation in Brazil and the United States*. Durham: Duke University Press.
Selka, S. 2007. *Religion and the Politics of Ethnic Identity in Bahia, Brazil*. Gainesville: University Press of Florida.
Sheriff, R. 2001. *Dreaming Equality: Colour, Race, and Racism in Urban Brazil*. New Brunswick: Rutgers University Press.
Skeggs, B. 1997. *Formations of Class and Gender: Becoming Respectable*. London: Sage Publications.
Skidmore, T.E. 1974. *Black into White: Race and Nationality in Brazilian Thought*. New York: Oxford University Press.
———. 1999. *Brazil: Five Centuries of Change*. New York: Oxford University Press.
Skinner, D. 2006. Racialized Futures: Biologism and the Changing Politics of Identity. *Social Studies of Science* 36 (3): 359–488.
Smith, C.A. 2016. *Afro Paradise: Blackness, Violence, and Performance in Brazil*. Urbana: University of Illinois Press.
Sovik, L. 2009. *Aqui Ninguém é Branco*. Rio de Janeiro: Aeroplano.
Stephan, N. 1991. *The Hour of Eugenics: Race, Gender, and Nation in Latin America*. Ithaca: Cornell University Press.
Telles, E. 2004. Race in Another America: The Significance of Skin Color in Brazil. Princeton and Oxford: Princeton University Press.
Twine, F.W. 1998. *Racism in a Racial Democracy*. New Brunswick: Rutgers University Press.
Wade, P. 2007. Race, Ethnicity and Nation: Perspectives from Kinship and Genetics. Ocford: Berghahn Books.
Wagley, C. 1957. Plantation America: A Culture Sphere. In *Caribbean Studies: A Symposium*, ed. Vera Rubin, 3–13. Seattle: University of Washington Press.
Winant, Howard. 1994. Racial Conditions: Politics, Theory, Comparisons. Minneapolis and London: University of Minnesota Press.
Wright, W. 1990. *Café con Leche: Race, Class and National Image in Venezuela*. Austin: University of Texas Press.

CHAPTER 4

Class Biographies in a Racialized World

I first met Bia a few days after arriving in Salvador in November 2005. I was staying with a friend in a quiet neighbourhood, and she invited me to the Goethe-Institut one evening—a beautiful architectural building where regular cultural events are held in the city centre. I was introduced to Bia, who at the time was absorbed with showing pictures of her wedding to a group of friends from her post-graduate psychology programme. Bia offered to help me to find more permanent accommodation. We developed a friendship and eventually I visited her home and met her childhood friend, Isabel. Bia and Isabel lived in the predominantly black working-class neighbourhood of Liberdade and both described themselves as belonging to the popular or working classes (*classe popular*). They were always busy, balancing their time between studying for their graduate degrees at public university, working part-time, and carrying out household responsibilities. Occasionally, they invited me to a family party or to stay overnight. On one such evening, they explained to me how challenging it was to improve their lives in Brazil. Bia and Isabel recounted that as young girls, they walked for miles to attend a Catholic school where several of the children were lighter-skinned and came from well-to-do families. After school, these children were picked up in a car by their parents while Bia and Isabel walked the long journey home. Once at home, they could not relax as they suspected their wealthier schoolmates did—there were chores to do, family obligations to fulfil, and school assignments to complete. They realized from an early age that to succeed, they would

D. J. Gordon, *Blackness and Social Mobility in Brazil*,
https://doi.org/10.1007/978-3-030-90765-5_4

have to work twice as hard to overcome the obstacles of living in a marginalized community.

Bia's mother—Norma, a divorcee—tried to take advantage of available educational opportunities for her children (a daughter, Bia, and a son, Danilo). Indeed, Bia successfully passed the challenging university entrance examinations to get into two public universities (a state and a federal university, to study for different degrees), and Danilo entered the police force. Bia and Isabel's friendship endured over time as they studied and took examinations together, worked on internships, graduated from university, and supported each other through major life events, such as career choices, buying property, marriage, and divorce. Fortunately, my friendship with Bia and Isabel has endured over the years as we exchange social media messages and pictures. I always see them on return trips to Brazil.

Today, the circumstances of Bia and Isabel's lives have changed significantly: they have moved out of their former community, circulate in more middle-class spaces, and have experienced considerable occupational mobility. However, they explained that they do not see themselves as proper middle class (*a classe média propria*), pointing out that their social life is rooted in the peripheral neighbourhoods where their family members and friends live. On a return trip in July 2017, they invited me to lunch at the iconic Yacht Clube da Bahia on the Avenida Sete de Setembro, a scenic avenue that runs along Salvador's coast. Isabel picked me up from my friend's apartment in her new sports utility vehicle. As we arrived at our destination, it was not lost on us that this used to be one of the most exclusive social clubs in Salvador, primarily geared towards the Euro-Bahian elite. Isabel, who is dark-skinned and wears her hair in a low Afro, was dressed in Black Power earrings and a colourful shorts jumpsuit. Bia arrived shortly afterwards with her mother, Norma, dressed in white lace and white linen blouses respectively, paired with jeans, bright lipstick, and high heels. We greeted and exchanged gifts, then selected a table in the restaurant near to a glass wall where we could look out at the stunning vista of the sailing boats and ships on the Bay of All Saints. Looking around at the mostly light-skinned patrons, they exclaimed that they had never eaten at this restaurant. The waiters seemed somewhat surprised but were polite and courteous as we ordered traditional Afro-Bahian cuisine—*moqueca de camarão* (a dish made with shrimp, coconut milk, tomatoes, peppers, onions, and palm oil)—and settled down to a lively conversation, catching up on the latest happenings in our lives.

While Bia and Isabel's choice to go to the Yacht Clube da Bahia for lunch perhaps symbolized their new class position and associates them with whiteness and privilege, they interact across class and race lines all the time, in all kinds of social contexts in the various corners of their lives. Bia and Isabel's stories of upward mobility are interesting in the sense that it illustrates that class position is sometimes not a dead end and that spaces exist for manoeuvring of status within a stratified and hierarchical society. Bia and Isabel are likely to be perceived as structurally and symbolically "white" by virtue of their social mobility and acquisition of "white" traits, including higher education and property in upscale neighbourhoods. Yet their stories challenge the popular ideas about the upwardly mobile black individual who eschews their family and friends as they become absorbed in the physical spaces and social networks appropriate to their changed class status. For example, although both Bia and Isabel enjoy hobbies that involve middle-class networks and require significant investment (Bia dances with a sword-wielding belly-dancing troupe while Isabel cycles around the city with a bicycle club) they also spend significant amounts of time in the largely peripheral, black working-class neighbourhoods where their relatives still live, visiting friends and neighbours and going to church. As such, they negotiate jarring realities daily, as do many residents of the city. At the same time, their stories are not necessarily representative of the small group of Afro-Brazilians discussed in this book. Obviously, people are not carbon copies of each other. Furthermore, people's socio-economic lives change over time in ways that inherently critique notions of class difference that only highlight the opposite ends of the socio-economic spectrum, that is, utter poverty or elitist privilege.

The people whose lives and voices are discussed here are not easily placed within well-defined class categorizations. They have different family histories, socio-economic realities, and future expectations. Like Bia and Isabel, the lives of many research participants seemed to curve around both sides of the working and middle classes. Indeed, quick and easy uses of class labels don't necessarily tell you all you need to know about a person's life—the same can also be said about race or any other social identity.

This chapter is organized as follows: first, I will discuss class theories and then go on to examine representations of "race," "class," and "gender" in Salvador, revealing some of the symbolic struggles that occur over the construction of social difference in a hierarchical and highly unequal society. Subsequently, I explore the contours of the black middle classes, giving insight into their diverse social, economic, and cultural origins, generational transformations over time, and a sense of the social processes underlying their social ascension. Finally, I draw on in-depth ethnographic material to present the trajectories of three of the families I studied, underscoring the diversity and heterogeneity in respondents' lives. There are stories of success over tremendous hurdles—exemplified by Bia and Isabel's trajectories outlined earlier—as well as stories of downward social mobility. The underside of these stories is the usually hidden workings of class advantage and disadvantage, complicated by race, gender, and other social identities, as people negotiated an unequal society.

Class Theories

Marxists have focused on "class" as the most salient aspect of an individual's socio-political identity.[1] From a Marxist perspective, systems of stratification derive from the relationship of social groups to the means of production. In conventional readings of Marxism, people's objective conditions are profoundly shaped by the nature of their relation to capital, to owners, and to their own alienated labour power. In a capitalist society, two distinct classes with diametrically opposed interests conflict with each other. Class struggle becomes the very motor driving social change. However, a social class does not necessarily see themselves as such (a class for itself) until it acquires class consciousness and solidarity, or a true awareness of the nature of class exploitation, which would inevitability lead to working-class revolution.

Some scholars have argued that this dualistic interpretation of societal struggle is problematic. For example, gender and race theorists have challenged these views, underscoring that gender, race, and ethnicity can't be reduced to class: indeed, they are important life-structuring variables in their own right.[2] Feminists have pointed out that some versions of Marxism ignore factors outside the traditional workforce as well as other gender-

[1] Useful literature include: John H. Goldthorpe (in collaboration with Catriona Llewellyn and Clive Payne), *Social Mobility and Class Structure in Modern Britain*; Anthony Giddens, *The Class Structures of Advanced Societies*; E.P. Thompson, *The Making of the English Working Class*; Raymond T. Smith, *Anthropology and the Concept of Class*; Eric Wolfe, *Europe and the People Without History*; John Roemer, *A General Theory of Exploitation and Class*; Nicos Poulantzas, *Classes in Contemporary Capitalism*; and Georg Lukàcs, *History and Class Consciousness*.

[2] For this type of argument, see Paul Gilroy, *There Ain't No Black in the Union Jack: The Cultural Politics of Race and Nation*; Cedric Robinson, *Black Marxism: The Making of the Black Radical Tradition;* and David Morley and Kuan-Hsing Chen, eds., *Stuart Hall: Critical Dialogues in Cultural Studies.* In terms of gendered aspects of class, see Beverley Skeggs' book, *Formations of Class and Gender.*

based realities that have important effects on peoples' everyday lives and life chances.[3] Other scholars have argued that Marxism does not capture the deeply fractured social worlds of post-industrial societies that we now inhabit, shaped by various transnational and global forces (McLennan 1979; Lipset and Marks 2000; Wright 1985).

Max Weber's multidimensional approach to class and his concepts of status groups, social honour, and prestige seemed most useful in understanding the complex urban context of contemporary Salvador, where it is difficult to place people into well-defined categories. Weber understood class as the weighted combination of a variety of different positional effects, as entailing both objective (such as income, occupation) and subjective factors (self-identity, subjectivity) and as arising in spheres of production and consumption (Weber [1947] 1964). In such a reading, other factors such as lifestyle (which includes cultural practices and skills) and status (which is more subjective, individualized, and variegated) are also important.

The French sociologist Bourdieu's approach to class is insightful here as well—especially his concept of "habitus," which refers to an external world of cultural assumptions and social institutions that ordinary people inhabit without thinking very much about them, and an internalized version of that world that becomes part of people's identities, generating dispositions to feel/think/act/judge in certain ways (Bourdieu 1978, 1990). In *Distinction* (1984), Bourdieu suggests a model of class based on capital movements through social space. The distribution of various forms of capital (economic, social, cultural, or symbolic) structures this social space and confers power and strength upon the holder. For example, masculinity and whiteness in Brazil are valued (and normalized) forms of cultural capital, constructed historically through struggles over assets and space. For Bourdieu, class is neither an essence nor an indeterminate set of fluctuating signifiers, but an arbitrarily imposed definition with real social effects (Moi 1991). Bourdieu's model enables an analysis of the micropolitics of power and the struggle between different groups for control of the social order. For example, my ethnographic data reveals that education, cultural knowledge, social networks, and bodily features are

[3] See for example Simone de Beauvoir, *The Second Sex: The Classic Manifesto of the Liberated Woman*; bell hooks, *Ain't I A Woman: Black Women and Feminism*; Gayatri Chakravorty Spivak, "Can the Subaltern Speak?" in *Marxism and the Interpretation of Culture*, ed. Cary Nelson and Stan Grossberg.

intimately connected and recognized as forms of symbolic capital in the negotiation of social position in Salvador.

As we shall see in the following paragraphs, peoples' actual life trajectories do not necessarily make sense in terms of being middle class or working class. Salvador is a complex urban environment, full of socio-economic diversity. Tremendous changes have occurred in the city in recent decades that have transformed the urban landscape and led to a realignment of social groups. For example, the traditional elite have reconfigured themselves and the informal sector has expanded to include a wide range of people of different socio-economic backgrounds. Therefore, in both this chapter and Chap. 5, I argue for an approach that factors in class-stratified interactions—both across and within classes. People have different realities and lifestyles—yet they may live side by side, work in the same place, or traverse the same public spaces.

Representations of Class

The contemporary "class structure" of Salvador has been studied by social scientists who have made it clear that it is undergoing a constant process of modification (Guimarães 1998; Castro and Sá Barreto 1998; Guimarães and Castro 1990; Guimarães et al. 1995; Agier 1994, 1995; Sansone 1992; Azevedo [1953] 1996). In Bahia, the formation of the petrochemical and offshoot industrial centres in Bahia increased job opportunities for middle-class professionals and created a new working class earning a reasonable salary plus important social benefits such as health care. Therefore, social mobility was an important aspect of the class structure until prolonged recession in the 1980s and 1990s. Several processes were sparked by these developments. For example, a spatial transformation occurred in that 50 per cent of the socially ascendant workers moved out of their neighbourhoods of origin to higher status, new *bairros* (neighbourhoods; Agier 1995). By the 1990s, researchers were pointing to an increased number of African descendants in middle-income occupations (Da Silva Soares 2004).

Despite these transformations, many research participants portrayed the urban-class structure in Salvador as immutable and unchanging, with sharp divisions between "us" (*o povo*) and "them" (*os ricos*; the lighter-skinned elite). Indeed, most research participants felt that Brazil was a racist society and argued that there were commonalities of oppression that joined poorer and wealthier blacks in racial solidarity. However, on other occasions they conceded to the existence of a small middle class and agreed that it was a reasonable description of their class position. They could also point to neighbourhoods they considered middle class; these locations were said to have organized streets, running water, good infrastructure, businesses, and shopping facilities. Some neighbourhoods were considered *nouveau riche*, where residents were said to drive imported vehicles and live in high-rise, luxury apartments. Older neighbourhoods could be viewed as rundown but still with middle-class segments. Figure 4.1 is a picture of Ladeira dos Aflitos, a slightly run-down area with large, older houses where the author lived for the longest period of time. Figure 4.2 is a view of luxury apartment buildings located in the Upper City along the Corredor de Vitória. Some of these buildings had trolley cars and piers to take residents down to the sea.

Generally, persons avoid discussing class directly in everyday conversation—rather, it is embedded in peoples' narratives about social encounters or situations. However, when prodded further in interviews, research participants explained Brazil's social structure primarily in terms of class, in

Fig. 4.1 Ladeira dos Aflitos, an older neighbourhood in Salvador. (Source: Doreen Gordon)

Fig. 4.2 Luxury apartments, Corredor de Vitória, Salvador. (Source: Doreen Gordon)

which a constellation of symbols such as neighbourhood residence, places where one socialized, lifestyle, acquisition of university degrees, types of leisure activities, and possession of private health insurance were important indicators of class status. However, there was also an acknowledgement of race as structuring opportunities for black people, locking them into poverty. For example, Jonas Figueiredo stated:

> Many people make the mistake that just because you can point to a successful black person, then others should be able to do it and that if they can't, it is because they are lazy. This is a gross misunderstanding of the situation. The

> obstacles for black people are difficult. Most of them are poor and for a poor person, it is virtually impossible to get into university with the bad quality of public school education these days. Private education is very expensive. Even if you get into public university where education is free, some courses carry a lot of expenses like Dentistry, Medicine, Law, and Journalism. Most people end up having to work and study. When you finally get your degree, you need the right contacts to get a job and rise in your profession. If you do become successful one day, it has as much to do with luck and connections as hard work and achievement. (Interview: April 21, 2006)

Here, Jonas not only pointed out issues of race in structuring black people's mobility but also the need for the "right" social connections and social networks in negotiating access to jobs and opportunities, or social capital.

Small, but important, differences along socio-economic continuums in Salvador exist. For example, fine distinctions were made about the type of house one lived in—such as whether it was owned or rented. Another important marker was whether a house was a self-constructed, unpainted one in a socially mixed area compared to a ready-made, modern apartment building/townhouse with a "proper" façade in a posh neighbourhood or gated community. One informant told me that you can tell it's the house of a middle-class person if they have more than one bathroom as well as a small room/toilet for the maid (typically in the service area or towards the back of the house). Indeed, homes were central sites of distinction through the use and organization of the space, presence of consumer and electronic goods, and the display of cultural investments (such as books, musical instruments, photographs, and paintings). For example, the Teixeiras' apartment was furnished with inherited antiques, a chandelier, paintings, and two beautifully carved wooden cupboards filled with fine crockery—testimony to their long-established status in the middle classes. This is perhaps different from the décor of someone newly arrived in the middle classes whose house may be furnished with more modern pieces.

There are also other, more intangible measures of social class, attesting to symbolic struggles of power between social groups—such as language (speaking proper Portuguese), tone of voice or accents, etiquette, body language, and tastes in music, leisure activities, and clothing. Pierre Bourdieu uses the term "cultural capital" to describe predispositions and sensibilities that one learns throughout life that tend to reproduce class status (Bourdieu 1973). Of the three elements of cultural capital, the notion of embodied capital suggests that the closer the people can come to presenting themselves as possessing and embodying predispositions and sensibilities of the dominant group (through tastes, accents, language, dress, and mannerisms) the more successful they will be in society (Bourdieu 1984). For example, I was invited to a party where I met the Teixeira sisters—Sandra, Anabela, and Anita. Friends had mentioned that they were an example of a "well-educated" black middle-class family. Dressed tastefully in fashionable clothes, Anita had her daughter, Isabela, with her. Isabela was about eight years old and was described as well-behaved and quiet—traits that were approved by friends and family as appropriate feminine characteristics. She was evidently aware of all the social rules, as she promptly informed me of my *faux pas* of pouring my coffee into the cups laid out for tea.

The Brazilian anthropologist Cecilia McCallum has noted that class in Brazil has often been represented as a static thing, rather than a social relation (2005, 101). In this sense, I am concerned with understanding the dynamics of daily interactions and the experiences of class in my respondents' daily lives. However, class positions are not just relative forms in social space—in some sense, they are institutionalized positions because the cultural capital of the middle classes (particularly the lighter-skinned middle classes) can offer substantial rewards in the labour market. Indeed, our social locations influence our movement and relations to other social positions and hence our ability to capitalize further on assets we already have. Thus, Brazilian researchers have shown that children of middle-class origins are more likely to arrive at middle-class destinations, hence limiting competition from the poorer majority, resulting in social closure and the reproduction of structural inequalities through the selective practices of recruitment to educational and occupational opportunities (Scalon 1999). For example, Túlio Santana, who was born into an established black middle-class family, was able to provide his children with even greater opportunities—such as high-quality education, privileged social networks, and exposure to international travel. This capital could be translated into greater social mobility, as two of his children have lived outside of Brazil (his son Martino is a musician living in Canada while his daughter Terrianna spent time living in Ireland). The downside of this is that symbolic struggles may become institutionalized through the provision of limited subject positions to inhabit for some Afro-Brazilians—an aspect of class dynamics that may constrain their lives in specific ways so that inequality is maintained. For example, black women, regardless of social background, constantly enter implicit trading arenas (such as carnival, annual festivals) where their sexuality, femininity, and respectability are judged in terms of value in which the rate is established by others.

In Brazil, social and racial prejudice are irrevocably linked (Roth-Gordon 2017). Thus, blackness is associated with a cluster of symbols linked to being lower class or poor, including *bêbado* (drunkard), *vagabundo* (bum), *desonesto* (dishonest), *mal educado* (uneducated/not knowing how to behave), *ladrão* (criminal/thief), *malandro* (rogue; up to no good), *puta* (slut/whore), *perua* (woman who likes to wear tight clothes), and *fedor* (bad small). These classed and raced stereotypes must be specifically re-worked by respondents in their daily lives, underscoring the racialized aspect of social mobility for Afro-descendants. For example, one informant, Paulo Ribeiro, a senior engineer in the state-owned oil company, Petrobras, liked to go out in carnival but underscored that one had to look good:

> You can't go out in any manner. You have to take a bath, put on deodorant, smell nice, look good, do your hair in a certain way. For the ladies, put on lipstick, wear nice clothes, and so on. You do this to avoid harassment and discrimination on the street. (Interview with Paulo Ribeiro, November 25, 2006)

Another informant, Jorge Carvalho, explained that he had to confront these racialized stereotypes when he first started working as a young man in a construction company. He described his success in negotiating these conditions and winning the admiration of his colleagues as "*minhas conquistas*" (my conquests). He stated that, "They could not deny that I was hard-working, disciplined, and that I got things done."

On the other hand, some research participants challenged these stereotypes. For example, blackness is associated with animalistic sexuality, implied in such words as *puta* (whore/slut) or *perua* (woman who likes to call attention to herself with tight clothes). I typically heard these terms among the social circles I socialized in whenever someone was referring to lower-class women's behaviour. As a member of a well-respected middle-class family for generations, Anita Teixeira (Helena Teixeira's youngest daughter) risked losing her social status by opting not to go to university and running off with an older man with whom she had a child. Her father and other family members were shocked, but she said that she was a rebel (*rebelde*) when she was young. When the relationship did not work out, she struggled for a while as a single mother until her mother asked her to return home so that Anita and her daughter could be protected by the family.

Gendered Representations

Middle-class black women had to confront and negotiate the strong symbolic associations between black women, poverty, and subservient positions, such as domestic work. Historically, black women in Brazil worked—under slavery they carried out hard physical labour alongside men in the sugarcane fields and in the plantation houses of the whites as domestic slaves. Up until the late twentieth century, the symbolic place of black women in Brazil was strongly associated with informalized domestic work—caring for the members of elite households as nannies, cooks, cleaners, and washerwomen (Medeiros 2018; Otovo 2016). However, as in other countries, rapid industrialization and rural-urban migration resulted in the reorganization of class structures and gender roles, including an increase in women's employment in the formal sector, thereby increasing their autonomy and independence (Besse 1996; Maier 2010; Sanabria 2016). Women entered factory work and service positions—for example, Miranda Santos' sister worked in a clothing factory in the interior of Bahia where she eventually rose to the middle-management position of floor supervisor. While women's movement into the formal labour force was met with some resistance by the Catholic Church and the state, throughout the twentieth century values and beliefs surrounding gender roles and the household continued to change and women's economic participation more than doubled (Edmonds 2010).

Beginning in the 2000s, black women's groups increasingly began to organize and react against traditional stereotypes of black women in Brazilian society, promoting a move out of the kitchen and into the public political sphere (McCallum 2005). They struggled to create a new set of symbolic associations for black women beyond the confines of traditional roles relegated to them, encouraging black women and girls to imagine themselves as doctors, lawyers, and other professionals. Furthermore, the media presented the success of some black women in politics as a sign of progress, especially of Benedita da Silva, the first black female senator, state governor, and federal minister in the history of Brazil. In another example, Didá Educational and Cultural Association of Salvador, created in 1993 by the percussionist Neguinho do Samba and now led by Cultural Director, Víviam de Jesus Queirós, stated that their main objective was to educate and empower black women and children through the arts and through educational programmes, thereby challenging traditional gender norms and constructs.

INTERVIEWER: What is the vision of Didá?

VÍVIAM: To realize that the black child and woman belongs to a very vulnerable group, that need to be prepared to take on social leadership, because the leadership will come from these women. Our vision is to empower and strengthen women but in a collective way to create social structures of getting along together and humanizing structures of power. We draw on the skills of black professors, activists and cultural performers who come in to give classes in music, percussion, dance, theatre, Afro-Brazilian history and culture, and feminist theories. So, there are free educational activities. We are also a carnival group and band. The idea is to increase black women's self-esteem, identity and citizenship through a focus on Afro-Brazilian culture. (Interview: June 29, 2017)

Clearly, this organization was working along with middle-class black activists and scholars who had a concern with promoting positive images of black female leadership. This could be interpreted as a class "project" of self-improvement, through education and appreciation of cultural/ethnic heritage. At the same time, Didá attempted to refashion previously denigrated types of work carried out predominantly by black women. For example, the 2007 carnival theme of Didá's group was "Samba Lavadeira" ("Samba of the Washerwoman"). More than 3000 women and children danced in bright blue costumes printed with images of washerwomen, dancing in front of the *trio eléctrico* (sound truck) that carried the Didá carnival band in order to celebrate and show appreciation for the contribution of these women to Brazilian society.

Mapping the Contours of the Black Middle Classes

Many of the older generation came from rural towns in Bahia. Often this generation had minimal education, though a few came from families where members once owned substantial property. They often migrated along networks of kinship, friendship, and patronage. These networks stretch across rural and urban areas and are maintained selectively, not least at milestone events such as infant baptisms, birthdays, weddings, and children's graduations. These relationships are often a source of support as well as information about matters such as jobs, housing, and schools. Yet the kind of ties of reciprocity and obligation that exist among rural migrants may be viewed negatively, since the drain on resources involved can impede a family's attempt to consolidate its wealth and move up the class structure.

Some were well established, urban families. Indeed, the names of traditional families still carry weight and prestige despite the rapid socio-economic changes that are transforming the city. A family considered to be "of good stock" is often referred to as having *sangue bom* (good blood), meaning that it could be trusted, having contributed to their community and having proved themselves over the generations as upstanding, moral, and respectable members of their society. Members of the Teixeira family were well-known educators who had brought up both related and unrelated children. The young inherited the friendship networks of their parents. This kind of cultural and social capital was vital for weathering difficult economic times, allowing them to draw on goods and services in kind to be repaid later.

A family's social status might be judged in terms of how well they had integrated into the middle classes through such traditional "whitening" strategies as acquiring a good education and respectable profession. At the same time, involvement in the black community through religious associations such as Catholic brotherhoods and *terreiros de Candomblé* (as well as funeral and burial societies) has traditionally generated different kinds of social prestige and connections.[4,5]

Occupational Trajectories

Teaching and education were strongly represented amongst my interviewees and has been a historically important avenue of social mobility, both upwards and downwards (Azevedo [1953] 1996; Dávila 2006). In one of the families I interviewed, the matriarch of the family, Helena Teixeira, was a retired history teacher in a public school. Her mother had been the

[4] For studies of religious brotherhoods and funeral societies in Salvador, see Julio Braga (1975), Russell-Wood (1974), Joao Reis (2003), and Kim Butler (1998).

[5] Earlier versions of these arguments were developed in a previous publication by the author. Please see: "Negotiating inequality: The Black Middle Classes in Salvador, Brazil," in *People, Money and Power in the Economic Crisis: Perspectives from the Global South*, 106–128, edited by Keith Hart and John Sharp (2014, Berghahn Press).

headmistress of a private school, while her second eldest daughter now teaches in a university in the interior of Bahia. Thus, teaching is a profession chosen by different generations in this family. However, it is less prevalent among the younger generation, since it is thought to be poorly remunerated and has lost some of its former cachet in comparison to other types of careers. Changing patterns of mobility reflect variations over time in occupational prestige.

In the past, many built respectable careers in middle-level government positions, were able to educate their children, and looked forward to retirement on a liveable pension. After the labour reforms of the Vargas government of the late 1930s and 1940s, public *concursos* (exams) became the normal way of acquiring a job within the expanding public sector. This degree of formality reduced the role of personal ties, social background, skin colour, and contacts in securing a job, thereby creating more opportunities for black Brazilians to acquire a stable job with employment benefits (De Santana 2009). As mentioned previously, the discovery of oil in the Bahian hinterlands, along with the establishment of secondary petrochemical and metallurgical industries, created a new labour market for skilled workers and employed many Afro-Bahians. However, this industrial sector had a lower multiplier effect on the Bahian economy than expected, as the investment it required reduced the scope for financing other types of industrial development (De Almeida 2006, 25). Since the 1990s, the public sector has contracted, and salaries have deteriorated in the face of privatization, deregulation, and decentralization as a result of neoliberal economic policies. Nevertheless, state bureaucracy has traditionally been a significant source of employment in Bahia and these jobs still have some prestige. The police force and the army, historically important sectors for people of African descent, continue to be viable career options (Butler 1998; Bacelar 1999, 2001). Rosa and her husband, Wesley Figueiredo, both originally from a small, rural Bahian town, managed to carve out a comfortable life living on Wesley's job in the police force, raising six children—Jacinda, Jussara, Antonio, Jonas, Rodrigo, and an adopted daughter, Marina. The eldest son, Antonio, patterned his father and went into the police force, while the youngest daughter, Jussara, works as a government administrator.

More recently, the services and financial sectors have become increasingly important sources of employment in the new neighbourhoods designed for Salvador's upwardly mobile professional middle class and for the booming tourist industry. During my fieldwork, Helena's youngest daughter, Anita, freelanced as an event organizer and insurance salesperson in these areas. But service sector jobs require intense interaction with clients, and this has increased the importance of *boa aparência* (having a good appearance). Traditionally, this meant having a more European physical appearance (lighter skin, straightened hair, slim build) while cultivating an air of sophistication and modernity through proper language use and personal style (Edmonds 2010; Roth-Gordon 2017). Some of my research participants said that it was not uncommon for employers to request photographs from job applicants, which they regarded as a discriminatory practice. It has even been suggested that heightened anxieties about appearance in the services sector may have even contributed to the increased popularity of plastic surgery in Brazil (Edmonds 2007).

Construction and transport are the areas where black people have tended to predominate. Traditionally viewed as jobs with low prestige, some participants nevertheless experienced substantial social mobility in these sectors. For example, Horacio Rocha worked as Chief Engineer of the train service in the Lower City and Túlio Santana was a senior engineer and administrator in a transport syndicate.

Some young people were making a success out of non-traditional careers, such as in the music and cultural industries. DJ Funk Soul had fashioned a career out of *música black* for over a decade and was poised to make a breakthrough in the industry. He has recorded with bands such as Negra Cor (Black Colour) and with well-known artists such as Daniela Mercury and Rebecca Matta. By combining the musical traditions of Africa, the United States, the Caribbean, and Brazil—including Reggae, Reggaeton, Afro-funk, Hip Hop, R&B, and Samba—DJ Funk Soul has positioned himself quite well in the commercial mainstream of black music and culture emerging in Salvador.[6]

Educational Trajectories

Education was significant to my research participants because it stood for equality in contrast to particulars such as colour. Thus, people of African descent could claim respectability and achieve social mobility through formal education despite the prevailing currents of racial discrimination.

[6] Please see earlier publication by the author: "Negotiating inequality: The Black Middle Classes in Salvador, Brazil," in *People, Money and Power in the Economic Crisis: Perspectives from the Global South*, 106–128, edited by Keith Hart and John Sharp (2014, Berghahn Press).

However, black children were often warned that they had to work twice as hard to achieve something. As one participant, Ana-Paula Humberto, said:

> My mother was lucky—she got a *bolsa* (scholarship) to send me to a private school. But there are children who don't realize what an opportunity they have and waste it. For this reason, they really need to be taught by their parents and others to understand their position in society, they need to be told the reality. You can't spend your time socializing with rich, white students who have money. If they fail their subjects, they will be alright. The poor black student won't get another chance. Parents need to let their children see this, rather than buying clothes and cellular phones so they can keep up with the fashions of rich students. Their money is better spent on education and basic needs. If I see my son watching Xuxa on television, I let him know that Xuxa has a child enrolled at the Pan-American school and that while he is watching TV, she is studying to get ahead.[7] I teach my children the reality. It does not mean I take away their childhood from them, but they also need to look critically at things and realize that they are not in the same position as privileged people, and that as Blacks they will have more obstacles to overcome. (Interview: December 15, 2006)

In Salvador's competitive job market, there is a growing premium on higher levels of education, so that younger people tend to be better educated than their parents. But some parents claimed that they had access to public education of higher quality. Since the 1980s, public education has markedly deteriorated along with the teaching profession's loss of status owing to poor wages and working conditions, declining morale and poor learning conditions (O'Dougherty 2002; Telles 2004). Accordingly, parents who benefitted from a good quality public education at school and university now must pay out much of their salaries for their children's private education to give them a chance of finding a good job. Horacio Rocha summed this up well:

> My father was a worker (*operário*) … a public servant (*funcionário público*). My mother … was a public servant as well. She only went to primary school. My grandparents were poor. But they had a different kind of poverty than exists today. It was a poverty that had possibilities for getting access to information, education. … I studied in a public school, and then a public university … it was possible until about the decade of the 80s when the public schools began to decline in quality. It was possible to get a good education.

[7] Xuxa is an extremely popular, blonde, and blue-eyed television presenter for children.

> In my street, all the *operarios* had access to public schools and universities and this made possible a certain social mobility, it was accessible. (Interview: May 24, 2006)

Many parents are obsessed with negotiating a child's entrance into a desired school. This involved a good deal of social networking with the aim of making direct contacts inside the school. There is a wide variety of such schools and they are informally ranked by the middle classes. Church schools, particularly Catholic and Baptist, have been a historically important and prestigious source of private education and some of them offer scholarships. The more modern and expensive private schools have smaller classes, and some offer an international, bilingual education.[8]

Brazilian education is unusual in that the best universities are publicly funded and attended overwhelmingly by children of the elite who could afford to give them the private schooling they need to pass the entrance examinations. Some of these private schools were derided with the expression *papai pagou passou* (literally translated, "papa paid so he/she passed"). Children from lower-class neighbourhoods, who attend increasingly demoralized public schools, generally attend private universities where they pay for inferior education. The people I knew often manoeuvred between private and public education at different times in their child's school career in the hope of their children passing the public university entrance examinations. In Rosa's family, the youngest son, Rodrigo, was shifted between public and private schools until he passed the entrance examinations for the Federal University of Bahia to pursue Journalism. While younger people do tend to be better educated than their parents, they also have higher expectations for job remuneration and satisfaction. They are more likely to perceive some work relations as authoritarian and to avoid what they consider degrading or humiliating conditions.

Income

The parent generation lived through the aftermath of World War II and then the dramatic socio-economic changes of the 1950s. During this period, public sector jobs were expanding, leading to more stable jobs

[8] Please see an earlier publication of the author, "Negotiating inequality: The Black Middle Classes in Salvador, Brazil," in *People, Money and Power in the Economic Crisis: Perspectives from the Global South*, 106–128, edited by Keith Hart and John Sharp (2014, Berghahn Press).

with benefits. Therefore, some of this parent generation had better economic conditions than their adult children today who operate in a labour market that prioritizes temporary contract jobs over secure work. Of course, not everyone shared in this period of greater prosperity. Nevertheless, people with salaries from formal employment and anyone who possessed a registered workbook (Carteira de Trabalho e Previdência Social) could count on benefits such as paid sick leave, unemployment benefit, and pension. The contractual nature of construction work, for example, meant that Pedro Santos spent periods of time unemployed; but because he had a registered workbook, he received unemployment payments that helped him to sustain his family. Pensions also provided some families with much-needed income—Helena Teixeira's pension as a retired teacher helped to support her youngest daughter and grandchild through economically difficult times. With stable salaries, some families were able to purchase property and better health insurance plans for themselves.

According to previous research (Figueiredo 2003; De Santana 2009), middle-class people of African descent have less wealth, savings, investments, and property than the more stable, lighter-skinned middle classes. Indeed, relying on cash from multiple sources of income, pooling the resources of family members, and juggling debts in a variety of forms (credit cards, supermarket discount cards, bank loans) were significant financial strategies for some of the families I knew when compared with the weight placed on savings and investment traditionally associated with the middle classes. As Angela Figueiredo has stated, "The black middle classes basically lives on their salaries, rather than on inheritance from their family … part of their income is spent on helping their relatives in need" (Figueiredo 2003, 256, *my translation*). She also pointed to black businessmen's lack of access to credit from established financial institutions, making it hard for them to maintain and expand their enterprises (Figueiredo 2003, 176). Her informants claimed that interest rates on loans were a deterrent to borrowing, and that, in any case, there was no one to act as an intermediary for them. According to her, their low levels of monetary and other kinds of capital, when combined with the discriminatory attitudes they encountered, posed great challenges for black business, ensuring that they remained vulnerable on the margins of the middle classes.

Other sources of state-based income were important to some families, such as cash-transfer programmes. Although Bolsa Familia (Family Stipend) and Fome Zero (Zero Hunger) programmes under Lula's administration were not the first targeted social programmes in Brazil's history, they framed more explicitly than before the idea of social security and protection as a right and public responsibility (Fenwick 2009, 110). Miranda's niece applied for and was accepted by the Bolsa Familia programme when she was living as a single parent in the interior of Bahia. She migrated to Salvador in 2006 with her two daughters and lived with Miranda for a few months. By 2013, the girls had completed secondary school and were preparing for the university entrance examinations, attaining a level of education higher than their mother.

Property

For my older interviewees, especially married couples, acquiring a house was extremely important and they went about it in different ways, drawing on the state, commerce, and social networks. Miranda and Pedro Santos, both in their 50s, bought their first home with a low interest loan from the Caixa Econômica—one of the largest government-owned financial institutions that catered mostly to a low- and middle-income clientele. Miranda also had property of her own in the interior of Bahia, inherited through her mother. At one point, she considered selling this land and leaving her marriage. The Teixeira family also inherited a house—along with other property and investments—from the mother's side of the family. They owned several apartments in Salvador, which they rented out for income. Helena Teixeira's daughters had a summer vacation house inherited from their father; other families pooled together to buy vacation property.[9]

Lifestyle and Attitudes Towards Consumption

Research participants emphasized that they did not live a lavish lifestyle, but thought that being able to visit the cinema, go to popular restaurants and bars, go on vacations, and having enough disposable income to buy fashionable clothes every now and again were important for their quality of life. Anabela Teixeira described a middle-class lifestyle as follows:

[9] Please see an earlier publication by the author: "Negotiating inequality: The Black Middle Classes in Salvador, Brazil." In *People, Money and Power in the Economic Crisis: Perspectives from the Global South*, 106–128, edited by Keith Hart and John Sharp (2014, Berghahn Press).

> You should be able to go to the cinema, the theatre, or a show—not much, but maybe once or twice per week. You would also have a car to get around, so you can travel to the good beaches … and you should be able to go to the mall … buy a couple books, though this is getting more expensive. (Interview: December 27, 2005)

Most respondents participated in annual events which sometimes entailed considerable expenses—such as carnival, religious festivals such as Lavagem do Bonfim, or São João festivities in the interior towns of Bahia. Expenses could involve rented accommodation, fashionable clothing, food, drink, tickets to enter events, and transportation. Others rented their apartments to foreigners travelling to Bahia for these events, charging in US currency. One informant managed to use this annual rent money to pay for her condominium fees as well as school fees, clothes, and shoes for her daughter. While carnival activities occurred in public spaces, it is a class-stratified event with distinct racialized tones. To avoid being like *pipoca* (popcorn) and jumping around in the streets (hence more vulnerable to harassment), people frequently chose to participate in *camarotes*—which were essentially party boxes that line the carnival circuit so that you can watch the carnival *blocos* (blocks or groups) and *trios eléctricos* (sound trucks with carnival bands). Each *camarote* offered a different package, along with clothing such as T-shirts. The most exclusive *camarotes* had services such as buffets and bars, air-conditioned lounges to relax, cyber rooms to connect to the internet, discotheques with DJs, and salons for styling the hair of patrons and keeping them glamorous. Amidst the festivities, however, black carnival *blocos* consistently complain that they are relegated to the less popular carnival circuits that do not get as much media coverage. Indeed, cameras zoom in and out of the crowd on the more prominent routes, where the *gente bonita* (the beautiful people; usually meaning lighter-skinned folk) vie for time in front of the camera.

Other respondents described a relatively quiet lifestyle, focused on socializing with family members and close friends at church, at Sunday dinners, and on special occasions such as birthdays and Christmas or Easter holidays. For example, Paulinho describes a childhood that was monitored closely by his parents:

> We were always supervised closely by family members. When my parents were not around, my grandparents came to care for us. We did not go out much to parties … but we had lots of friends from the neighbourhood and

from school. So, we visited people in their homes. We did not have a car, even as young adults. That is because we did not need to socialize much outside of the house or outside of the family. (Interview: January 3, 2006)

Some informants who lived in low-income neighbourhoods had more disposable salary and savings than others living in more high-income, fashionable areas of the city who were struggling to make ends meet and to keep up with the consumption patterns of their class. Martino (son of Túlio Santana), a university student studying music and living with his parents, once said:

> There are a lot of people in this neighbourhood who are slaves to capitalism. The middle class wants to spend more and more, and accumulate more and more, only the more you spend, the less you have. There is no way of saving your money, it is a vicious cycle … credit cards, cheques, bank loans … these are the things of the middle class. Sometimes we see the case of a person who prefers to have a fancy, brand-name car instead of remodelling his apartment. This is the thinking of the middle class: they want to show that they have it, but the truth is that they do not. They want to buy this and that, but they cannot pay for it. This generates a lot of social discrimination—as much on the part of middle-class people as between the lower classes and the upper classes. It is a vicious circle and I don't see an end to it unless people stop and think about their role in their communities and families, their responsibility to society and, more importantly, to themselves. (Interview: March 17, 2007)

This quote highlights an important variant in how to be middle class. Martino explicitly rejects the excesses of a neoliberal market society while advancing something different from the materialistic values espoused by a rising bourgeoisie. It is a critique of the particular way of being middle class associated with a modern culture of consumerism and credit, as distinct from the more traditional forms of consumption and marking status. This may differentiate older and newer black middle classes, as de Aguiar Costa Pinto (1998) noted.

The Teixeiras, as representatives of an older black middle class, were one of the first families on their street to acquire electricity, a fridge, and a telephone. This was how they announced their status as modern Brazilian citizens. They also put more emphasis on education and cultural experiences (such as foreign travel, visits to museums and exhibits) than they felt some of their neighbours did. One of Helena's daughters, Anabela, found a particular neighbour ridiculous because she put her child in the most fashionable clothes instead of investing in educational and cultural activities for her child. Anabela shared that this same child, who was going to a private school that was located near to a public school, objected to wearing the same colour uniform blouse as the public school. Nita thought it was unfortunate that the child seemed to be learning the mother's values, as her family had no money to speak of. Thus, she was providing a critique of her neighbour's consumption values as she felt this family was in a more precarious class situation than herself. The Teixeira family's own relative downward mobility (being stationary when so many others were moving up) was evidently irritating. Their response was to hold ever more firmly unto a vision of education and cultural enlightenment as the foundation of middle-class position.[10]

Family Trajectories

I met Rodrigo Figueiredo, a journalist in his early 30s, in the first few months of fieldwork. I subsequently got to know many of his relatives and friends. Those close to him described him as calm, shy, and friendly. His salary as a journalist for a major newspaper company in Salvador allowed him to leave his family home at the age of 27 and live comfortably by himself in an apartment. He subsequently moved in with his girlfriend in an apartment on a quiet street in a neighbourhood close to a university campus. Rodrigo's parents had migrated from a small town in the interior of Bahia and had managed, through the father's stable government job as a musician in the police force, to build a house near to the city centre and all the modern services offered there. Rodrigo described his parent's neighbourhood as follows:

> The neighbourhood was *bem misturado* (very mixed), built on a *ladeira* (steep slope). At the top were the more middle-class families, with the more expensive houses. The houses at the bottom of the slope were simpler, mostly self-constructed. My father bought land here and built his house. It was not designed or planned. (Interview: December 27, 2005)

[10] Please see an earlier publication by the author: "Negotiating inequality: The Black Middle Classes in Salvador, Brazil." In *People, Money and Power in the Economic Crisis: Perspectives from the Global South*, 106–128, edited by Keith Hart and John Sharp (2014, Berghahn Press).

Rodrigo was the youngest child of his parents and grew up in a time when they were more economically stable. As a result, he had greater opportunities for education than his siblings and was the only child to have completed university. After graduating, he spent some time living in the United States before returning to Salvador. At this time, he related that he was only one of four Afro-Brazilian journalists in the newspaper company he worked for. Later, he completed a Master's in International Relations in Argentina. More recently, he was offered a writing fellowship in Germany where he stayed for a couple of months.

As a young man growing up, Rodrigo remembers receiving a good allowance from his parents, which enabled him to buy nice clothes, go to the mall with his friends and watch a movie, or go on weekend retreats with his church peers. However, he remembers a time during the 1970s and 1980s when things were not as good financially—for example, there was no money for entertainment or clothes, just for basic household expenses. Yet they did not feel that they were poor—they ate well and all the children were educated up to secondary school level.

Rodrigo emphasized that all his siblings were socialized to be independent and to work hard, so he did not have to support any of them. He does not own an apartment or a car, yet he considers himself middle class because of where he lives, his education, his occupation as a journalist, his travels abroad, and his lifestyle. He is involved in varied social networks because of his job. For example, he knew the popular Afro-Brazilian actor Lázaro Ramos just before he became a celebrity.

While his family initially defined themselves as black middle class, this was not always flagged as important in their daily practice. One evening I was relaxing with them after a family dinner I had been invited to and we were discussing their understandings of race and class in Salvador. They claimed a strong black identity and described incidents of racism which they had experienced living in the city. Furthermore, they joked that even the lighter-skinned members in their family were black because they grew up around black people. However, as I got to know the family over time, their religious and working identities appeared to be more important matters in everyday life. Family members belonged to an evangelical church for many years. This church did not have an established building—rather members met in small groups in each other's homes. However, they networked with similar groups, creating a network of believers across Salvador and other cities. Regular activities were often planned during the week, including Sundays and holidays—such as Bible readings, prayer meetings, retreats, summer camps, young people's gatherings, games nights, seminars, or workshops—which provided opportunities for members to socialize and receive support and care. While he is no longer involved in this group, Rodrigo saw this experience as a foundational one, where he learnt important values and interpersonal skills. Today, he still maintains friendships from this group, which included persons of different social backgrounds and skin colour.

The Figueiredo family's religious views shaped their views about race and class in Salvador. For example, family members tended to distance themselves from racial politics, arguing against voting on the basis of race or ethnicity. They often derided the introduction into public universities of racial quotas as a kind of philanthropy. Rather, they supported the racial democracy ideology and the view that all people are equal before the law and before God. However, they were very enthusiastic about Rodrigo's relationship with a European woman and tried to encourage that relationship as much as possible as an additional symbol of his mobility into higher social circles.

The Humberto family had a different orientation to the Figueiredo family. This family had lived in Salvador for generations and were respected members of their community. The matriarch of the family, Maria, lived in Garibaldi, a traditionally black working-class neighbourhood. Also living with her was her son Rafael, a highly educated engineer who had been trained in the United States. Her eldest son, Silvio—a tall, dreadlocked man in his 40s—lived in an apartment in an upscale neighbourhood surrounded by mostly lighter-skinned neighbours. He worked with the government as an economist, was a politician, and ran his own non-governmental organization dedicated to preparing poor, black students attending public schools to take the challenging examinations for entrance into university. Maria's daughter, Ana-Paula, lived on the floor above her with her husband and two children. She worked along with her brother Silvio in his organization. Through their work, they were in touch with powerful local and national figures as well as intellectuals and activists across Latin America and the United States.

Maria, in her 70s and surrounded by her family who have done well in life, is very attached to her house. It was where she got married, where she had her children, and where her grandchildren are now being raised. Over the years, she and her husband had made significant improvements to the house; therefore, she does not want to move out of the neighbourhood. Her children pooled their resources and acquired an apartment building a few blocks down the road for their mother. She manages the building and the rent contributes to her household income.

In comparison to most residents of Garibaldi, this family is perceived as having a higher status because family members are embedded within complex community, national and international networks that have elevated their position, both in the eyes of their neighbourhood and in the larger urban community. Maria recounted that she was born among the *classe popular*, in a rented house where her father was a fisherman and her mother a housewife. She had very basic education and her parents had none. However, today, her children are university educated, the eldest one with a doctoral degree from the University of Campinas (in São Paulo state). She remarked that the family had always been racially conscious, because her father used to work as a butler in white people's houses and swore that he did not want his children to perform humiliating work for whites. All of them completed primary-level education—however, Maria had to start working to help support her family. Today, all of them have their own house through their hard-working efforts, she emphasized.

In contrast to both the Figueiredos and the Humbertos, the Teixeira family considered themselves to be long-established middle-class, privileged equals of the Euro-Brazilian elite for generations. I first met members of this family at a friend's birthday party in a *bairro fino* (posh neighbourhood). The Teixeiras stood out in a crowd of mostly light-skinned guests. Three sisters of around the same age, they were dressed elegantly, with expensive looking handbags and shoes, well-groomed hair, and slim bodies with smooth, brown complexions. My Brazilian friends whispered to me, "They are different from the common person here in Salvador. You can tell because they are *bem educado*. Look at how beautiful they are!"[11]

Amongst their friends and acquaintances, the Teixeira family was not regarded as being typical in a city where most residents are poor and visibly darker-skinned, and where the higher social classes are symbolically linked to whiteness. According to Helena (the sisters' mother), her family were descendants of an original union between a slave woman and the son of a wealthy Dutch sugar-plantation owner. Her Dutch ancestor subsequently married a white woman, but all the children from the former union were well taken care of. Helena's parents lived a comfortable life

[11] This means that a person is "well-educated" in terms of formal schooling, but also in manners, etiquette, and cultural knowledge.

and were respected in their neighbourhood—her father was university educated, worked for a British shipping company, and spoke many languages, while her mother was a teacher. Helena said that in past generations, they would not have seen themselves as black. They often referred to themselves as *mulata clara* (light-skinned, mixed race). She recalled as a child hearing her relatives make disparaging remarks about the physical features and moral traits of dark-skinned people. Such discourses on appearance and the value of particular physical traits no doubt possibly helped non-white elites to distance themselves from blackness (and poverty) as a strategy of avoiding racial discrimination themselves, although this did not always succeed. Helena told me that many years ago her mother, Rita, tried to enter the exclusive social club, Yacht Clube de Bahia. Rita had many long-standing, influential friends among the Euro-Brazilian elite who frequented the club. One of them was the *prefeito* (mayor) of Salvador and he wrote a letter recommending Rita as a member. Even so, the executive committee of the club rejected Rita's application. Her friends were indignant and were convinced that Rita's skin colour was the deciding factor. To protest the club's decision, they immediately left it. This story reveals the character of everyday social relations in several ways. First, it indicates how personal relationships across lines of colour could sometimes mitigate the negative effects of having dark skin, although it did not always protect them from discrimination. Second, Rita's friends might not have perceived her or treated her as black—she had become structurally and symbolically white by virtue of being a member of their class. Unlike previous generations, however, the younger members of the Teixeira family now claim a more politicized, black identity—even if it increases the chance of encountering discrimination.

The Teixeira family once lived in high-status neighbourhoods with grand houses and many servants. They no longer have this kind of lifestyle and have fallen on harder times. However, they are not poor and have managed to slow down their descent on the social scale. Thus, they moved from a large elegant house to a smaller, more efficient, and affordable apartment—but it is located in a fashionable part of the city, close to a popular swimming beach and tourist area. Through selective consumption of items of cultural value and astute social networking, the Teixeiras manage to maintain their middle-class status, thereby distinguishing themselves from a blackness that is still associated with poverty.[12]

These three cases of upward and downward mobility provide a more complex picture of the black middle classes than is normally available in the media and academic literature. Rather than revealing a stable, fixed lifestyle defined by occupation, income, or consumption, the cases above show people moving unsystematically between different social levels. They also give a sense of the different strategies employed to deal with racial inequality and discrimination, including family and community ties, friendship with the lighter-skinned elite, getting an education and profession, and asserting or denying the existence of racism in particular contexts. Indeed, the middle classes are Janus-faced—at one moment identifying with the social system, at others going against it while identifying with poorer people (and with blackness). An interesting theme is the stories told about the Yacht Clube de Bahia by respondents—the encounters that some families had with the exclusionary and discriminatory practices of the club in the past are juxtaposed with Bia and Isabel's entrance into its exclusive spaces to have lunch generations later, attesting to racial transformations that have occurred in Salvador. The analysis of class in this chapter aimed to capture the fuzzy boundaries of race and class in a modern, Urban City, rather than to fix it in place.

References

Agier, M. 1994. 'Classe' ou 'Raca?' Socialização, Trabalho e Identidades Opcionais. *Bahia: Análise e Dados, O Negro* 3 (4): Marcha.

———. 1995. Racism, Culture, and Black Identity in Brazil. *Bulletin of Latin American Research* 14 (3): 245–264.

Bacelar, J. 1999. Black Brazil. In *Black Brazil: Culture, Identity, and Social Mobilization*, ed. L. Crook and R. Johnson. Los Angeles: UCLA Latin American Center Publications, University of California.

———. 2001. *A Hierarquia das Raças: Negros e Brancos em Salvador*. Rio de Janeiro: Pallas.

[12] Earlier versions of these arguments were developed in previous publications by the author. Please see: "A Beleza abre Portas: Beauty and the Racialized Body among Black Middle-Class Women in Salvador, Brazil" (July 17, 2013, Feminist Theory 14: 203-- 218); and "Negotiating inequality: The Black Middle Classes in Salvador, Brazil," in *People, Money and Power in the Economic Crisis: Perspectives from the Global South*, 106–128, edited by Keith Hart and John Sharp. (2014, Berghahn Press).

Besse, S.K. 1996. *Restructuring Patriarchy: The Modernization of Gender Inequality in Brazil, 1914–1940*. Chapel Hill: University of North Carolina Press.

Bourdieu, P. 1973. Cultural Reproduction and Social Reproduction. In *Knowledge, Education, and Cultural Change*, 71–84. London: Tavistock Publications.

———. 1978. Sport and Social Class. *Social Science Information* 17: 819–840.

———. 1984. *Distinctions: A Social Critique of the Judgement of Taste*. Cambridge: Harvard University Press.

———. 1990. Structures, Habitus, Practices. In *The Logic of Practice*, 52–65. Cambridge: Polity.

Braga, J.S. 1975. Sociedade Protetora dos Desvalidos: A Irmandade de Cor. MA Dissertation, Universidade Federal da Bahia, Salvador.

Butler, K. 1998. *Freedoms Given, Freedoms Won: Afro-Brazilians in Post-Abolition São Paulo and Salvador*. New Brunswick: Rutgers University Press.

Castro, N.A., and V. Sá Barreto. 1998. *Trabalho e Desigualdade Raciais: Negros e Brancos no Mercado de Trabalho em Salvador*. São Paulo: Annablume.

Costa Pinto, Luiz. [1953] 1998. *O Negro no Rio de Janeiro: Relações de Raça Numa Sociedade em Mudança*. Rio de Janeiro: UFRJ.

Da Silva Soares, R. 2004. Negros de Classe Média em São Paulo: Estilo de Vida e Identidade Negra. PhD Dissertation. Universidade de São Paulo, São Paulo.

Dávila, J. 2006. Diploma de Brancura: Politico Social e Racial no Brasil, 1917-1945. São Paulo: Editora UNESP.

De Almeida, P.H. 2006. A Economia de Salvador e a Formação de sua Região Metropolitana. In *Como Anda Salvador e Sua Região Metropolitana*, ed. Inaiá Carvalho and Gilberto Corsa Perreira. Salvador: Editora Universidade Federal da Bahia.

De Azevedo, T. [1953] 1996. *As Elites de Cor numa Cidade Brasileira*. Salvador: Editoria da Universidade Federal da Bahia.

De Beauvoir, S. 1974. *The Second Sex: The Classic Manifesto of The Liberated Woman*. New York: Vintage.

De Santana, I. 2009. À Margem do Centro: Ascensão Social e Processos Identitários Entre Negros de Alto Escalão: O Caso de Salvador. PhD Dissertation. Universidade Federal da Bahia, Salvador.

Edmonds, A. 2007. The Poor Have the Right to Be beautiful: Cosmetic Surgery in Neoliberal Brazil. *Journal of the Royal Anthropological Institute* 13: 363–381.

———. 2010. *Pretty Modern: Beauty, Sex, and Plastic Surgery in Brazil*. Durham: Duke University Press.

Fenwick, T.B. 2009. Avoiding Governors: The Success of Bolsa Familia. *Latin American Research Review* 44 (1): 102–131.

Figueiredo, A. 2003. A Classe Media não vai ao Paraíso: Trajetórias, Perfis e Negritude entre Os Empresarios Negros. PhD Dissertation. Universidade do Rio de Janeiro, Rio de Janeiro.

Giddens, A. 1973. *The Class Structure of Advanced Societies*. New York: Harper and Row.
Gilroy, P. 1985. *There Ain't No Black in the Union Jack: The Cultural Politics of Race and Nation*. Chicago: University of Chicago Press.
Goldthorpe, J.H. (in collaboration with Catriona Llewellyn and Clive Payne). 1987. *Social Mobility and Class Structure in Modern Britain*. Clarendon Press.
Guimarães, A.S.A. 1998. *Um Sonho de Classe: Trabalhadores e Formação de Classe Na Bahia dos Anos Oitenta*. São Paulo: Editora Hucitec.
Guimarães, A.S.A., and N. Castro. 1990. Classes, Regimes Fabris e Mudança Social No Nordeste Brasileiro. *Cadernos* 12: 11–32. Centro de Recursos Humanos.
Guimarães, A.S.A., M. Agier, and N. Castro. 1995. *Imagens e Identidades do Trabalho*. São Paulo: Editora Hucitec/ORSTOM.
hooks, b. 1984. *Ain't I A Woman: Black Women and Feminism*. Boston: South End.
Lipset, Seymour M. and Gary Marks. 2000. It Didn't Happen Here: Why Socialism Failed in the United States. New York: Norton.
Lukäcs, G. 1971. *History and Class Consciousness*. Cambridge: MIT Press.
Maier, E. 2010. Concluding Reflections: Renegotiating Gender in Latin America and the Caribbean. In *Women's Activism in Latin America and the Caribbean: Engendering Social Justice, Democratizing Citizenship*, ed. Elizabeth Maier and Nathalie Lebon, 337–351. New Brunswick: Rutgers University Press.
McCallum, C. 2005. Racialized Bodies, Naturalized Classes: Moving Through the City of Salvador da Bahia. American Ethnologist 32 (1): 100–117.
McLennan, David. 1979. Marxism After Marx: An Introduction. Boston: Houghton Mifflin.
Medeiros, M.A. 2018. *Marriage, Divorce, and Distress in Northeast Brazil: Black Women's Perspectives on Love, Respect, and Kinship*. New Brunswick: Rutgers University Press.
Moi, T. 1991. Appropriating Bourdieu: Feminist Theory and Pierre Bourdieu's Sociology of Culture. *New Literary History* 22: 1017–1049.
Morley, D., and Kuan-Hsing Chen, eds. 1996. *Stuart Hall: Critical Dialogues in Cultural Studies*. London: Routledge.
O'Dougherty, M. 2002. *Consumption Intensified: The Politics of Middle Class Daily Life in Brazil*. Durham: Duke University Press.
Otovo, O.T. 2016. *Progressive Mothers, Better Babies: Race, Health, and the State in Brazil, 1850–1945*. Austin: University of Texas Press.
Poulantzas, N. 1974. *Classes in Contemporary Capitalism*. London: New Left.
Reis, J.J. 2003. *Death is a Festival: Funeral Rites and Rebellion in Nineteenth-Century Brazil*. Chapel Hill: University of North Carolina Press.
Robinson, C. 1983. *Black Marxism: The Making of A Black Radical Tradition*. Chapel Hill: University of North Carolina Press.

Roemer, E.A. 1982. *General Theory of Exploitation of Class*. Cambridge: Harvard University Press.

Roth-Gordon, J. 2017. *Race and the Brazilian Body: Blackness, Whiteness, and Everyday Language in Rio de Janeiro*. California: University of California Press.

Russell-Wood, A.J.R. 1974. Black and Mulatto Brotherhoods in Colonial Brazil: A Study of Collective Behaviour. *Hispanic American Historical Review* 54 (4): 567–602.

Sanabria, E. 2016. *Plastic Bodies: Sex Hormones and Menstrual Suppression in Brazil*. Durham: Duke University Press.

Sansone, L. 1992. Cor, Classe e Modernidade em Duas Áreas da Bahia: Algumas Primeiras Impressões. *Estudos Afro-Asiáticos* 22: 143–174.

Scalon, C.M. 1999. *Mobilidade Social No Brasil: Padrões e Tendências*. Rio de Janeiro: Editora Revan.

Skeggs, B. 1997. *Formations of Class and Gender*. London: Sage Publications Ltd.

Smith, R.T. 1984. Anthropology and the Concept of Class. *Annual Review of Anthropology* 13: 467–494.

Spivak, G. 1988. Can the Subaltern Speak? In *Marxism and The Interpretation of Culture*, ed. Cary Nelson and Stan Crossberg. Urbana: University of Illinois Press.

Telles, E.E. 2004. *Race in Another America: The Significance of Skin Color in Brazil*. Princeton: Princeton University Press.

Thompson, E.P. 1963. *The Making of The English Working Class*. New York: Pantheon.

Weber, M. 1964 [1947]. *The Theory of Social and Economic Organization*. New York: The Free Press.

Wolf, Eric. 2010. *Europe and the People Without History*. Oakland: University of California Press.

Wright, E.O. 1985. *Classes*. London: Verso.

CHAPTER 5

Negotiating Sameness and Difference

In this chapter, I show that class is central to how inequalities are lived, consolidated, reproduced, or rejected by respondents. Class did not just inform people's subjectivities, it also shaped power relations and symbolic struggles between different social groups over resources and opportunities. These struggles are racialized in ways that have specific implications for socially ascendant black people but also raises questions about their involvement in the black "community" and role in challenging these inequalities.

Drawing on ethnographic material, I will explore relationships across lines of class and race—especially the contentious relationship between middle-class and lower-class black people and critiques of black leadership. Middle-class black people both identify with and draw distinctions between themselves and *o povo* (the masses), symbolically associated with blackness. Thus, the material presented in this chapter is about the experience of class (or how class is lived) and its intersections with other social identities (e.g., race, gender, religion). In every judgement that respondents made about themselves, a measurement was made against others. In other words, the designated "other" (the masses/the people/the popular classes) was constructed as what they were *not*. However, at other times, research participants asserted their belonging to the black "community."

D. J. Gordon, *Blackness and Social Mobility in Brazil*,
https://doi.org/10.1007/978-3-030-90765-5_5

Lived Experiences of Class in Salvador

Class in the form of invidious distinctions of money, status, and privilege was both noticed and felt as significant. One common factor that heightened the awareness of class included going to a school associated with higher status (as in the case of Bia and Isabel); undergoing a crisis in the family such as death or divorce; or moving into a new neighbourhood.

> **Nubia:** When my vegetarian restaurant started to get good business, we moved into a new neighbourhood. We were surrounded by white people. I was now divorced from my second husband, who was white, so I did not have his protection or money. I was alone with my daughter, who was having trouble adjusting to her new private school. You know, she experienced a lot of racism, especially regarding her physical appearance. The streets were so quiet and organized in the area, you felt that if you had a big party, it would be a huge problem. (Interview: March 19, 2006)

> **DJ Funk Soul:** When my mother died, I felt like it was the end of the world. I was only 19 (years old). I was finishing up my degree in physical education, which my mother was encouraging me to do before she died. She didn't really support the deejaying and thought I needed a degree that would provide me with a steady income. When she died of cancer I was devastated. My uncle helped me to finish my degree and I was fortunate to inherit this apartment from my mother. Then I started working in the days at a school as a physical education teacher. At nights I would go out and play at different events. It was hard becoming an adult without my mother. My aunties and grandmother offered to help me clean and cook, but I told them no … occasionally, I would allow them to. I had to build my name as a deejay. The music business is very competitive, and the equipment is not cheap. Also, you must get to know celebrities and shit. It takes time to begin to earn any money from it, but I was persistent. I am only seeing the results now, after about 10 years of just working, working. (Interview: May 14, 2006)

Research participants traversed different areas of the city for various reasons and often crossed the barriers of categories such as "shanty towns" or "gated-communities." Nevertheless, these movements tend to be selective—research participants did not feel equally comfortable in Salvador's different neighbourhoods nor did they always have the resources to gain access to exclusive areas of entertainment. For example, DJ Funk Soul explained that:

> I do not walk around in neighbourhoods where I do not know anyone. If I know someone living there, that is fine. I once went to the suburbs, but I don't socialize there. I go very little to places like Bairro de Paz or Ribeira. I have some reluctance to go there, because I do not know the area. However, it's a question of time as well. I do not have anything against these places ... but you should have a reason why you are going there because there is a lot of crime. You can't just arrive there, looking around and walking around aimlessly. It might be easier for women to enter these places than men. But I do not have prejudices against these areas. I mean, all cities are like this, you must be careful. (Interview: May 14, 2006)

He went on further to state that he did not have a car and had no issues taking public transport to get to his various social engagements—however, public transport hampered his access to various social spaces.

> The social life here is cool. There is the beach and lots of entertainment—carnival and parties throughout the year. But there are not many options at night. On top of that, everything closes early. Also, the public transport is very bad. You can wait at a bus stop for an hour and a half. So, I often end up having to rely on taxis to get around and that can get expensive. (Interview: May 14, 2006)

Among my research participants, there was the perception that certain neighbourhoods were associated with high levels of violence, such as parts of Liberdade or the peripheral areas of Salvador. However, other informants pointed out that upscale neighbourhoods were not safe either.

Several informants had the experience of living in different neighbourhoods over the course of their lives. For example, Paulo was born and grew up in the Cidade Baixa (Lower City), in a neighbourhood with a lot of factory workers, which he described as a mixed area.

> To me, all the rich people lived on top, in the Upper City. In the Lower City ... in my neighbourhood ... there was a mixture of people where I lived ... there were black people, white people, mixed people. Some people lived in houses, others in apartments and some even lived in what they call *palafitas* (houses raised on stilts). But sometimes even people who lived in the *palafitas* had televisions. There was a lot of pollution in the Lower City. This part of the city is associated with a lower-income population. It's a commercial area, with many factories, so people had a lot of health problems such as respiratory problems. But it was still a good place to live, there were various groups of people and things to do, including going to the sea. I

> studied in a public school which in my time, had a good standard of education; a mixture of people went there. I lived in an apartment with my parents and did not have a sense of want. My father had a car. I had friends of different social background and I was a very social person, involved in clubs and activities in school. I was exposed to a lot of situations and different kinds of people—prostitutes, drug users, gays. I knew them all, although I never took drugs. First, I went to live in Liberdade in an apartment. After getting a job in Petrobras, I went to live in Iapi, in an apartment in a popular neighbourhood. I met a woman who moved in with me. I continued to study and take *concursos* and moved up in the company. When I got married, I bought a house in Stella Marris. (Interview: November 25, 2006)

Thus, we can see that while respondents lived in varied neighbourhoods across the city and over their lives, issues of location, positioning, and movement through social space and place affected their access to and experiences of certain social spaces in the unequal urban landscape of Salvador.

Hidden Injuries of Class in a Racialized City

Representations of class are not straightforwardly reproduced but are resisted and transfigured in their daily enactment. They operate not only as an organizing principle which enables access to and limitations on social movement and interaction but are also reproduced at the intimate level as a "structure of feeling" (Williams 1961, 1977) in which doubt, anxiety, and fear inform the production of subjectivity—for example, when someone feels "out of place" or "slighted" (Moreno Figueroa 2008). This experience underscores the ways in which power and prestige are "felt" in interpersonal relationships and in different spheres of life. Following Ortner (2003), I refer to this experience as the "hidden injuries of class."

Some informants were more sensitive than others to the exclusionary practices of class. In other words, who "sees" and acknowledges class tend to be persons lower down on the social hierarchy; people higher up were more likely not to "see" it. For example, Martino, who lives with his family in the modern, middle-income neighbourhood of Ondina, reported that:

> We (his family) can't conceive of separating persons because of the colour of their skin, between a black or white person we don't see much difference, everybody is a human being. So, we don't see the reason for separating the black person from the white person or the *mestiço* … nothing has ever

> happened to me when I am out with my girlfriend who is white. Her family does not discriminate. No-one I know has a problem and I never suffered from any type of discrimination, either on the street, at a restaurant, in a bar, or at the cinema. As an inter-racial couple, we are not treated differently. We walk together. But there are people who will think that, because the majority of black people are poor in Salvador, a black man must be with a white woman because he is taking advantage of her money or she is a tourist who is looking for a black man to spend her summer holiday with. (Interview: March 16, 2007)

On the other hand, Bia expressed feeling discriminated against at different points in her life. The most obvious settings in which this occurred were at the private school her mother sent her to, at the public universities where she pursued tertiary education, and in her work settings. In school, she remembers feeling disgusted at the unfair social differences among students. At the time when I met Bia, she was pursuing her Masters' degree in Psychology at the Federal University of Bahia and working part-time as a Development Analyst. At the university and in her work-place, she met a wider circle of acquaintances, including white middle-class people and foreigners. She also was invited out to restaurants, bars, and parties by her peers. When I asked her about these friendships, she explained that:

> Although there is no segregation in Salvador, you feel like you are being treated differently, you feel how people respond to you when you are at work and in the university. I am in contact with people who have less than me and people who have more. When I am in a posh place, I immediately feel like … this is not my world. Many times, I feel discriminated, you feel people judging you, with comments on your clothes, hair, or where you live. Sometimes it is not even something they say to you directly, they are just having a normal conversation and forget that I am there. They assume some things, like access to transportation. They all have their cars to get home safely at night. I have to arrange transportation with my brother because I live far away, and the buses are infrequent at night. None of them at the restaurant are going to take me home because they are afraid of where I live. (Interview: March 18, 2007)

However, Paulo Ribeiro, who had grown up in the Lower City among the *classe popular* (popular classes), declared that he rarely felt discriminated against:

> I have never been much discriminated against even though I grew up in the Lower City. I mean, there are these small things, like who has the straighter nose or the straighter hair. This was more of a high school thing, where the girls wanted the guys with the pretty, straighter hair. I remember going to the barber and wanting him to cut my hair like one of those boys with straight hair, but my hair just could not fit into the style. I made a scandal in the barber's shop because of it. I never experienced discrimination because I have always been a person able to move between different environments … to make connections with different kinds of people. (Interview: May 25, 2006)

It is possible that Bia and Paulo's perspectives may have differed because of gender—Paulo may have had more freedom of movement and opportunity to socialize as a man, while Bia's life had been more constrained and limited to her family, neighbourhood, and church, which are legitimate and respectable spheres of socialization for many black women (Burdick 1998).

These hidden injuries of class are difficult to define because informants considered it highly impolite to talk about the economic situation of other people or themselves. Soteropolitanos would assert their belief in the universality and dignity of all people, in equal access to public goods, and a comfortable standard of living for all. In practice, however, people's actions were contradictory. For example, one evening Bia was showing pictures of her wedding to her university peers. They praised the wedding dress, the baroque architecture of the Catholic Church in Nazaré, and teased her about her handsome, brown-skinned groom, who was grinning from ear to ear in the photos. Of similar class background, this relationship was approved and supported among her friends and family. At the same time, a white American friend of theirs, who was also completing the same post-graduate programme, began to date a part-time security guard at the orphanage where she volunteered her time. As the relationship developed and became more serious, the reaction was one of deep horror and concern from her Brazilian friends. One of them remarked that while she knew American women often had affairs with their gardeners and personal trainers, in Brazil these kinds of boundaries were rarely crossed because the difference between them was too wide. This example suggests that while relationships between persons of the same social position were not censured, intimate relationships between persons of different social locations were. Friendships could be negotiated, but the idea of intimacy across sharp class divides often made people uncomfortable.

Friendships Across Lines of Class and Race

Ties of friendship are meaningful for overcoming challenges and obstacles in life and for deriving affection, emotional comfort, and feelings of belonging. Although friendship is perceived among my informants as a matter of choice, individual preference, and affiliation, in fact most persons felt at ease with friends of their own class or social background. To avoid social embarrassment and uncomfortable situations, such as not being able to afford the prices of restaurant food, most people stuck to their own social milieu. In some instances, friendships were inherited from previous generations of family friends. For example, Helena Teixeira regularly met up with long-time friends who knew her family well, some of whom are from the Euro-Bahian elite. They share pictures of their travels, children's events, and so on. They may also arrange shopping trips or go on month-long vacations with each other to other parts of Brazil or Latin America. These friend's children are close to Helena's daughters as well. However, the Teixeiras, known for being open-minded and cosmopolitan, also incorporated new friends into their network—such as migrants from other part of Brazil, foreigners, and people of different sexual orientation.

While Salvador is a highly stratified urban society, people do develop a variety of different levels of friendship and ways of relating and identifying with each other in different social contexts that convey feelings of trust, empathy, and affection without necessarily dissolving social hierarchies. The casual and warm way in which people normally greet each other (such as hugging, a hand on the shoulder, a handshake, and a kiss on the cheek) implies an equality between them, even though this may not be the case. Rezende (2002), in her study of friendships among middle-class people in Rio de Janeiro and London, presents a notion of the Brazilian person in which ideas of individuality and choice co-exist with the need for relating to others. She suggests grappling with the notion of friendship in Brazil by looking at it as an idiom of affinity rather than one of equality (1994).

Bia and Isabel both had friendships that had been forged outside of their neighbourhood—in their church, university, work, and recreational spaces where they met people of different social backgrounds. However, when I quizzed Bia about the friendships in her post-graduate programme at university, she underscored that she did not know how permanent these friendships were.

> Let's see if these friends last after we are finished [our programme]. Now Monica, she is really a nice, caring, cool person. I like her a lot ... she is a good person. She has been to my house in Liberdade and I have been to her apartment in Campo Grande. But I do not know if we are close just because our studies bring us together and we have a common interest. Most of the time, I hang out with friends from here. As you can see, I am not that accustomed to going to the places that they like to go to, and I can't stay out late at night because of transport issues back to my neighbourhood. So, I do not know if we will continue to be friends. (Interview: March 15, 2006)

Nevertheless, these "higher-status" friends had been invited to Bia's wedding and they often spoke highly of her as someone they liked and greatly admired, as someone who fought hard for what she wanted and believed in. They desired to have more contact with Bia and invited her to many events, some of which she attended. Bia, however, was cautious about becoming too close. Isidora, who was from this group of university friends and considered Bia a friend, lived in Brotas with her mother in a large house with a backyard and had access to a car. She stressed that most of her friendships were with women and usually of the same age and class. She spoke about her friendship with Bia—and was aware of their class differences. She acknowledged that her parents were able to help her financially and that she had had experiences such as foreign travel that Bia had not been exposed to. She knew that Bia had to work and look for scholarships and loans. She thought Bia's life was more stressful compared to her own. She also knew that Bia had problems with transport, especially if they went out on Saturday night. Therefore, they mostly socialized on the university campus doing projects together, going somewhere to eat lunch, or studying together to complete an assignment. Isidora stated that Bia was very pleasant, cared for her family, and was a sensible and practical person. She stated that she had invited Bia to her house in Brotas to use her internet to complete assignments—but that she had never been invited to Bia's house in Liberdade. She hoped that they would remain friends after graduation, though Isidora has since migrated to Portugal.

Genuine friendships across lines of class and race do develop that endure the test of time, such as the relationship between Rodrigo and Marcia. Rodrigo met Marcia—a light-skinned, blonde, and blue-eyed young woman—when he was active in the youth group of the Evangelical church they attended. The church organized activities for the young people, such as retreats, movie nights, and games nights. Both Marcia and Rodrigo referred to this period in their life as important in the formation of their character, where they learned to respect other people's differences and love themselves and each other. She emphasized that as a group, they socialized with each other intensely. Though neither is still involved in the church, they have remained good friends through adult life, even as Marcia has migrated to Germany where she works as an artist. Rodrigo has since been to Germany where he visited with Marcia—and Marcia has visited Salvador several times. She acknowledged that there was racism in Brazil and that she knew that there were some places where the security guards would look twice at someone like Rodrigo. She explained that given her skin colour and other physical features, she would get in easily at the popular night clubs, for example, but that the security would probably pause at Rodrigo unless he was with her. She joked that if Rodrigo could not get in, then the poor black man who did not have the "right" clothes or friends did not have a chance.

Another example of a friendship across colour and class lines is the relationship between Paulo Ribeiro and Marco who both work for the same company, Petrobras. Marco is from an upper middle-class white Bahian family and lives in an apartment in Pituba. He trained as an engineer and met Paulo at Petrobras, where Paulo was one of the few dark-skinned employees in a senior management position. Paulo and Marco were both divorced; only Paulo had adult children. They could not have been more opposite personalities with different life experiences—but their friendship has endured over time. Marco explained that his social circles were very narrow before he met Paulo. He described himself as quiet and having lived a sheltered life. On the other hand, Paulo is very social and talkative, funny, and liked to host parties. Their friendship developed slowly as they gained an affinity and admiration for each other. They recognized that they shared similar values, so they started socializing with other friends from work in various locations around the city. As the trust between the two deepened, they eventually decided to be business partners and opened an ice cream shop in the Lower City. This business did well for a while—however, they closed it down when it was no longer profitable. Marco admires Paulo's work ethic and ability to overcome many challenges in his life, including racism, to reach his present position. He often encouraged Paulo to try various *concursos* that Petrobras had available and supported him in his endeavours. It is not clear though, whether Paulo and Marco would have been such good friends if Paulo was still working class—as one moves up the social ladder, class interests, perspectives, and values change.

Making Distinctions

Among older generations, education and work are particularly important markers of distinction. Having a regular job and being successful at it took on added meaning for my research participants, especially because of the association of people from the Northeast with being lazy, hedonistic, and undisciplined. Education and work take on significant dimensions for socially ascendant African descendants because engagement with work, educational, and occupational success and material reward are important for offsetting the effects of skin colour and establishing one's position as a respectable citizen. Occupational success and material wealth are valued in an environment that supposedly functions on a meritocratic and democratic basis as opposed to ascriptive forms of evaluation.

In Brazil, as in other parts of the world, the concept of work has undergone several transformations. As Medeiros (2018) has argued, the word has long been associated with manual labour, servitude, and poverty, so that a classed and raced division exists between those who carry out manual work (symbolically associated with black and mixed-raced populations) and those whose work is more intellectual (symbolically associated with white populations). However, once slavery was abolished in 1888, the land-holding elite, in need of workers, tried to break down these stereotypes and encourage people to work, including the construction of training schools in the rural areas to educate the poor and the introduction of legislation to limit semi-nomadic and subsistence lifestyles (Medeiros 2018; Mayblin 2010). As a result of some of these efforts (as well as other cultural ideas and practices), participation in and commitment to wage labour slowly acquired a moral value—although this was not universally adopted or uncontested (Silva Diniz 1991; Mayblin 2010). Therefore, Tarcis—a lawyer in his 70s and step-brother of Pedro Santos—once affirmed that, "*Sou negro de moral!*" (I am a black man of moral principles), while referring to the discipline and hard work that it took to acquire his profession. In claiming this identity, Tarcis simultaneously distanced himself from the negative stereotypes associated with blackness, poverty, and manual labour through his work identity as someone carrying out intellectual labour.

However, some research participants positively identified with their parents' categorization as a manual or industrial labourer and took a similar path. Paulinho, whom I met at a wedding, was a computer engineer in his 30s. His family lived in Garcia, a predominantly black neighbourhood in Salvador. He explained that his father was a mechanic, but not formally trained. Paulinho described his father as a *trabalhador* (worker) who carried out hard, physical labour. He exerted a lot of pressure on his children to pursue their education but also encouraged them to acquire a technical skill. Paulinho and his brother went to a technical high school that aimed to prepare them for industry—Paulinho as a computer engineer and his brother, Raul, as an electrical engineer. In this case, being a *trabalhador* and working for industry was reworked as something positive from one generation to the next.

The legitimizing ideology of education pervades all social classes in Brazil, as education represents progress, modernization, and democratic values, as well as being in touch with the cultured world. It was a path through which Afro-Brazilians could claim respectability and achieve social mobility, by drawing lines between the educated population and the illiterate masses, symbolically associated with blackness. Indeed, an ideology of education exists that suggests that one class is educated and mannerly, while the other is uneducated and ignorant. Thus, educated black people could demarcate themselves from the kind of uneducated blackness associated with poverty.

Educational credentials can be converted into economic, social, and cultural capital and had whitening and modernizing effects (Dávila 2006), as well as being a value upon which achievement-oriented assessments of the individual could be made. However, educational institutions in Brazil have played and continue to play a leading role as instruments of division. Despite the general deterioration in public school education, there were still some schools regarded as providing a good education. However, the private schools have more variety in educational tracks and options as well as more individualized attention with smaller classroom sizes. They also played a role in the formation of social and political networks in the upper middle classes and elites of the city (Borges 1992). There were also schools attached to the Baptist and Catholic churches, as well as secular, non-denominational schools that promoted creative and innovative educational methods and options. However, parents at times expressed concerns about the homogeneity of these schools in terms of class and race and the narrow experiences they afforded. Horacio Rocha recounted going to his daughter's elementary school to complain about the portrayal of black people in a history book that she was reading. He expressed his views on their education as follows:

INTERVIEWER: Where did your children go to school?

HORACIO: All of them went to private school, because I think there has been a degradation of public schools. The quality of these private schools is good. They studied at Génesis for their primary education and Cândido de Portinari for their secondary education.

INTERVIEWER: Where are these schools located?

HORACIO: In Pítuba and Costa Azul. But these schools are very homogenous, very middle-class, you know?

INTERVIEWER: Were you satisfied with this?

HORACIO: No. But my children liked it. One of the schools, Génesis, was more liberal and encouraged individual expression and creativity and my daughter enjoyed that experience. But to balance the very middle-class environment of their school I used to try and expose them to different experiences, different parts of the city.

INTERVIEWER: Like what kind of experiences?

HORACIO: They went to church with us as children. Up to today, I try to get them to think critically about their society. I take them to poorer neighbourhoods in the city. I try to get them interested in voluntary and community work.

INTERVIEWER: Do you think it has worked?

HORACIO: (Laughing) I think my children are more middle-class than I am. But they grew up with the kind of family values that were instilled in me. They occasionally come with me to church and they take the bus like everyone else. (Interview: May 24, 2006)

Thus, some parents who were able to afford private education for their children in the hopes of placing them within a better social position felt ambivalent about these options as their children became more distanced from the parent's social background and from blackness. I will now move on to look at symbolic practices of distinction between middle-class and lower-class black people, starting with an ethnographic vignette from my fieldnotes about spending the holidays with one of the families involved in this research.

(DIS)IDENTIFICATIONS OF CLASS

I was invited to join the Teixeira family for the New Year's Eve celebrations on the island of Itaparica. I met the three sisters—Sandra, Anabela, and Anita—their boyfriends, and some of their close friends at the harbour in the Lower City, where the catamarans and small boats departing for the islands off the coast were stationed. Meanwhile, Helena Teixeira and her brother Bruno took the family car and went across the harbour by ferry. Women stood in long lines, dressed in bikini tops, shorts, and skirts; men carried large containers of food and drink. *Pagode* music—a popular, speeded up version of the older *Pagode Romântico* (a sub-genre of Samba)—belted out from the radio of cars, taxis, and buses parked nearby. Some bare-footed children passed by selling peanuts, beer, and ice-cold bottled water to people waiting in the hot sun for departure. If one wanted to find where the black middle class was, one could observe them going on vacation weekends to the islands, the family dog on the back seat of the car. The island of Itaparica used to be more upscale, attracting the wealthy as well as a few foreigners who built their dream houses along the coast in quiet fishermen's villages. However, as the wealthy migrated to the recently developed and exclusive resorts along the north coast of Bahia, called the Linha Verde, those with less capital moved in. Itaparica therefore became a popular vacation destination for Soteropolitanos, who would rent rooms in local houses or who perhaps had a vacation house on the island, as was the case with the Teixeiras. Figure 5.1 shows a coastal area on the island of Itaparica, a frequent spot for vacationers.

We were lucky enough to get a ticket on the more comfortable catamaran ahead of the lines of people waiting outside, because Anabela had found someone she knew who could *dar um jeitinho* (give us a little favour). As we gazed at the scenery of the city getting smaller and smaller, Anabela filled me in on family news. Her mother, niece, and uncle were already on the island preparing food for the evening meal. As we arrived at the port on the small island and breathed the scent of the bean cakes frying, however, we could not resist stopping at an *acarajé* stall, where a large Baiana woman generously spread dried shrimp, chopped green tomatoes, *vatapá*, and pepper unto this traditional street food.[1] We then boarded a

[1] *Baiana* women are autonomous women who work as street vendors and typically sell traditional Afro-Bahian food such as *acarajé* and *abará*. *Acarajé* is a food that comes directly from West African food traditions: *they are* small fried balls made from milled black-eyed peas, known as "fradinho" in Brazil. The fritters are usually opened and served with condi-

Fig. 5.1 Island of Itaparica, off the Coast of Salvador. (Source: Doreen Gordon)

crowded minibus, heading towards a small village where their beach house was located. Later that night, we took a stroll around the coastal community. This quiet corner of the island gradually transformed throughout the evening as hundreds of people arrived in cars and on motor bikes, blaring Brazilian Popular Music (Música Popular Brasileira, or MPB) on their stereos. In some cases, car trunks and doors were opened to reveal a sophisticated sound system inside, creating the atmosphere of an energetic street party. Young girls with long dyed blonde hair sat on the backs of motorbikes in shorts and tight tops. Young men walked around in long beach

ments such as green tomatoes, peppers, dried shrimp, and *vatapá* (a dish made from bread, shrimp, coconut milk, finely ground peanuts, and palm oil mashed into a creamy paste). *Abará* is a popular Brazilian snack consisting of black-eyed peas or beans that are mashed into a paste, then steamed in a banana leaf. During the period of slavery (1500–1888), the trade of *acarajé* helped many enslaved women to buy their freedom (Torres 2007). Until nowadays, whole families are sustained with the trade of *acarajé* and other food products. Since 1992, *baianas de acarajé* have been united in an association that represents their interest as street vendors. Since its creation, the *Associação de Baianas de Acarajé e Mingau* (Association of Baianas of Acarajé and Mingau) has won the right to work as autonomous workers in the city of Salvador, Bahia.

shorts and flip flops, showing off their gym-toned muscles and looking slightly dazed, drinking beer. Anabela and her friends shuddered. "Look at what this island is coming to!" they said, or "*aqui e muito misturado*" (it is very mixed here), or "*o povo chegou*" (the people/masses have arrived). We spent the rest of the night trying to find a bar tucked away from the crowd and the music.

The following morning, we went shopping for the midday meal, looking for a local fisherman who could sell us some fresh crabmeat. I was going to learn how to make a favourite Bahian dish, *moqueca de caranguejo*, or crab meat simmered in coconut milk, palm oil, tomatoes, spices, and pepper. The fisherman looked quite bright, saying that he was getting good business over the holiday. But unfortunately, he said, the crowds also attracted the criminal elements from the city, as some people had already been robbed at gunpoint. My hostesses shook their heads, fearing that the island was degrading rapidly and that their once quiet get-a-way was becoming less safe.

Before cooking the meal, we headed off to a beach bar to have a couple of drinks, sarongs tied fashionably around our waists, distinguishing us as respectable women amongst the crowds. As we sat down for a beer, we noticed a few couples dancing *Arrocha* music at the bar next to ours. *Arrocha* is a music/dance style said to originate in the rural areas of Bahia, associated with less prestigious styles of music such as *Seresta* and *Brega*. While watching a couple dancing closely together, they exclaimed, "Look at how they are dancing!" Indeed, the woman was lifting her leg occasionally, tightly wrapped in the arms of her partner, enjoying the music and the sensual rhythm. This did little to endear her to the group I was with. "My goodness, it is so vulgar, she is lifting her leg up like a dog," one woman commented, holding up her dark glasses and scrutinizing the couple more closely. The group I was with discussed how this style of music was very degrading of women, that it was commercial popular music that had no style or well-thought-out lyrics to it.

In these last few sentences, claims to respectable female behaviour, regulated sexuality, manners, and appropriate public behaviour could all be utilized as practices of distinction in making middle class selves. These values were purportedly taught within the home and constituted important ideas about proper socialization encompassed by the term *educação* (education). As Borges (1992) has argued, the notion of "educated behaviour," or "manners," evolved in another era, a colonial era in which legitimations of position were couched, if not in racial terms, then in terms of socialization. This was a hallmark of a class on the move, a class asserting its difference from the masses while grasping power: a class that was non-white but educated and therefore qualitatively different from the uneducated. In the case of middle-class black women, downplaying their sexuality in a cultural environment where blackness is often associated with uninhibited and uncontrolled sexual instincts and behaviour might seem particularly important, as Salvador's thriving sex and tourism industry draws upon representations of the sexualized black woman. Distinguishing oneself from these representations of black female sexuality, middle-class women could align themselves with respectability and higher moral values.[2]

Ideas about proper moral behaviour were also revealed in the attitudes of some informants to extra-marital affairs and children born outside of marriage, although attitudes are changing, and respondents expressed varying opinions and interpretations on the practice. Traditionally, from a middle-class point of view, lower-class family practices were associated with children born out of wedlock, seen as the product of unregulated behaviour, or immoral behaviour. Unregulated procreation means large families and insufficient means to acquire the material resources and privacy of a middle-class household. Consequently, children are believed to reach maturity in an undisciplined environment. According to Ana-Paula, who distinguished her family from other types of families in her own and surrounding popular neighbourhoods:

> My children can't learn anything from their maid, so I don't leave them in her company for too long. You see them here in the neighbourhood. These women don't have any *educação* (education). They have children with different men and then expect their mothers to look after them. This is irresponsible. Do you think I could leave my children with my mother? All they are interested in is working to save enough money to enjoy themselves at carnival and São Joao. (Interview: March 12, 2007)

As Ann Stoler has argued in her article about sexual and racial frontiers in the construction of European and colonial identities in Southeast Asia, education and upbringing emerged as national projects at the turn of the century in both metropole and colony (1992, 518). In a similar way, anxieties over national identities and the process of *mestiçagem* in Brazil

[2] Aspects of these arguments were presented in an earlier publication by the author. Please see: "A Beleza abre Portas: Beauty and the Racialized Body among Black Middle-Class Women in Salvador, Brazil" (July 17, 2013, *Feminist Theory* 14: 203–218).

focused on the regulation of public health, hygiene, education, sexual morality, and moral upbringing. Children's upbringing and the family were crucial sites in which future subjects are seen to be made. As Otovo (2016) has shown, the domestic lives of Bahians—especially women and children—became increasingly subject to public scrutiny by a wide range of public and private institutions charged with improving the Brazilian population as they policed the moral borderlands of the community, both of its marginal and of its full-fledged members. At the heart of this process of inclusion and exclusionary practices, was a search for essences that joined formulations of national and racial identities.

Climbing the social ladder is not only a matter of gaining positive goods—for example, a better house, a nicer neighbourhood, more possessions—but also drawing negative lines between oneself or one's family and those below them. Such drawing of lines may be considered irrational, invested with fears of social pollution and danger (Douglas 2002). But this irrationality and emotionality can be understood in part as part of the anxieties of social mobility itself: the fact that such mobility is often tenuous, especially for the black middle classes and especially in the beginning; the fact that it carries ambivalent feelings of leaving kin, culture, and even pieces of the self behind; the fact that the socially ascendant self may feel out of place in the higher class surroundings; and more.

Some middle-class African descendants lived in proximity with lower-class African descendants and socialized in some of the same settings. The tendency was that as soon as they were financially able, socially mobile Afro-Brazilians moved out into a higher-status, middle-income neighbourhoods, associated with whiteness, modernity, and cosmopolitan lifestyles. However, this was not always the case. There are well-established musicians and some lawyers and doctors who stay in the working-class communities in which they grew up. Elite members of black cultural organizations also tend to be based in working-class communities.

The tension between upwardly mobile African descendants and their neighbourhood of origin was a particular characteristic of social mobility for black people. Persons who realized their obligations to family were deemed *pessoas de amor* (people of love). When Rebhum (1999) asked her informants in the northeast what love meant, she got responses that were more about one's relationship to the wider community rather than references to deep personal affection, intimacy, or sexual attraction. The kinds of responses she received focused on honest business practices, interpersonal respect, public obligations to support and protect the weak, and courtesy in general (Rebhun 1999). Moral judgements of character can become racialized in the sense that an upwardly mobile black person who is seen to have cut most of his former ties or refuses to honour family obligations may be accused of having "whitened" and therefore of thinking too much of himself/herself, of forgetting his/her roots, and of becoming arrogant and vain.

A person who is knowledgeable is a person who is *sabido*—but having too much knowledge can make one proud (*orgulhoso*), or too self-aware. The greatest compliment that can be given to someone in a prominent position is that they are *uma pessoa sem orgulho* (they are a person without pride). More educated persons are expected to act morally—having a sense of one's relationship to other people as a particular kind of sociality. This was particularly strong in the working-class neighbourhoods in which some informants live. Ignoring this sociality could lead to accusations against the upwardly mobile black person who is perceived as attempting to distance themselves from blackness. This is the dilemma which socially ascendant individuals must negotiate. Ana-Paula recalls that her family always practiced good, neighbourly relations.

> My father always was good to our neighbours and he passed that on to us. If he drove past a bus stop and saw a neighbour waiting, he would stop his car and give them a lift home. Things like that. Now that I work for the Steve Biko Institute, I often look for children in this community who could benefit from such a programme. (Interview: March 12, 2007)

However, Ana-Paula and her family often felt vulnerable to the envy and the possibility of "evil eye" (*mal olhado*) from their neighbours. When I lived with Miranda Santos for a short while in the initial months of fieldwork, she had explained to me the concept of *mal olhado*. She said that sometimes there were people in the community who might not even have the intention of hurting you but through their envy of that person's success, they could make them ill. The only cure was to enlist the help of a *rezador* or a Candomblé priest or priestess. Ana-Paula noted that:

> I think that people in this community think that we like to imagine ourselves as better off than them. Although we have been living here a long time and people know us, we also keep to ourselves and stay close. I think that is part of the problem: we don't mix with them on the streets. We tell everything to our family. We stay in the house and do things together. I don't keep a lot of friends. I'm close to my immediate family. (Interview: March 12, 2007)

Similarly, Ana-Paula's sister, Marcia—who lived a few streets down the road with her husband—did not like living in the community because of the antagonistic feelings she felt were directed against her higher status and economic position. She said she could sense the "evil eye" of the *povo* (people).

Partly as a result of these tensions and partly because of the need to define themselves as different to their neighbours, members of Miranda Santos' family engaged in practices of distinction—for example, the children and grandchildren stayed inside of the house. They could invite friends over but were not allowed to visit neighbour's houses. They were not allowed on the street—that realm of disorder and danger so well described by Da Matta (1991)—unless accompanied by an adult or older sibling for a specific reason. Anxieties about downward mobility into the surrounding, disorganized masses were indirectly expressed through family member's description of the changes that their neighbourhood had undergone in the past couple of years, regarded as both positive and negative. They referred to the increasing size of their neighbourhood, higher unemployment rates, increased levels of violence (due to organized crime gangs, police, and security forces and everyday crimes such as robbery, assault, and domestic violence), and a rise in impersonal social relations as some of the negative changes. Some family members no longer felt safe and feared encroachment of insecurity and criminality in their neighbourhoods and in their day-to-day living as they made their way across the city.

Fears of declining quality of life and deterioration of their neighbourhoods by the invasions of the poor were often expressed by research participants. For example, Marycelia—Miranda's niece—feared that her sons would become influenced by the drug gangs and popular culture of their neighbourhood. They had had many arguments, as the mother felt that her sons liked *a vida da rua* (the life of the street). DJ Funk Soul also remembers how his grandfather, with whom he lived for a while when he was younger, enforced strict rules about him coming home from school directly to study for several hours in the afternoon, therefore keeping him off the streets of the surrounding communities.

Yet at the same time, some informants expressed powerful needs to belong to and identify with the neighbourhoods and families from which they came and to show their commitment to the community. Often the present day, low socio-economic position of African descendants was linked to history and their past exploitation as slaves. In a conversation with one of my informants—a university professor—about the visual presence of large numbers of black and brown people in the poorer areas of the city, she explained that:

> Salvador has many blacks. And why are these blacks poor? If we look at history, the abolition of slavery was recent—1888. In truth if we think about the history of Brazil, the period of slavery represents a much longer time than the period after it. With the abolition of slavery, what was Brazil going to do with so many blacks that were once slaves? He had always worked in the house of someone and as a result was in the street with nowhere to go, without employment, without education … and so it continues until today. One hundred years is not enough for change, to transform this picture of poverty and exclusion. It is difficult to change this, especially because in the past there was no political directive to change the unequal situation of blacks. (Interview: July 5, 2006)

These explanations about the low socio-economic position of blacks were often framed within a historical discourse of slavery that linked the de-cultured slave (meaning that he/she was robbed of history and ethnic pride, exploited, and then subsequently excluded from education) with the present-day conditions of the working class. This could be interpreted as a kind of moral discourse about the condition of lower-class blacks from which they needed to be "rescued." Consequently, some informants were involved in different kinds of projects aimed at improving the conditions of poor, black, and mixed-race people. It is important to note that these efforts were not always welcomed by lower-class black people.

Community Involvement

"Community" was a contradictory term for research participants as they aligned themselves in multiple, ambiguous, and sometimes contradictory ways to their localities and to various organizations, including Black Movement and cultural organizations as well as religious groups, revealing varied class and race "projects." I employ the term "Afro-Brazilian community" in two senses here: first, to refer to residential localities where people of African descent are concentrated, such as when residents use the term "*bairro de negro*" to refer to places like Liberdade and Garcia. Second, I also use the term to refer to communities of practice based around self-consciously Afro-Brazilian political or cultural activities. Salvador has been a focus of study for its rich Afro-Brazilian culture, a term that implies that there is a well-bounded and stable set of cultural traditions and practices that people in Bahia share (De Santana Pinho 2004). However, the black "community" is itself a symbolic construct upon which people draw, rhetorically and strategically, by those claiming to do something for or in the best interest of the community (Cohen 1985). However, community is not just a "mental" construct—it is as much a product of "thinking" as it is of "doing." In other words, it is also what people do together—what they discuss, their participation in civil society organizations or events associated with Afro-Brazilian traditions and practices, and how they assess each other's performance of community that produces a sense of belonging.

According to Kim Butler, since the early twentieth century, Afro-Brazilians have employed ethnicity as a strategy of social and political advancement, especially as they were denied that power in the societies that they lived (Butler 1998). However, in Salvador, different groups assign radically different meanings to traditional symbols and practices associated with Afro-Brazilian culture and identity. For example, Candomblé practitioners are associated with Afro-Brazilian culture, black consciousness, and resistance while Evangelicals are often equated with being anti-social and intolerant of Afro-Brazilian cultural forms. However, it is complicated because there are some Candomblé leaders who do not participate in racial politics and there are some Evangelicals who do. Thus, there are multiple, overlapping, and competing *communities*, underscoring the heterogeneity of Afro-Brazilian identity politics.

Some research participants became activists involved in non-governmental organizations, seeking to add skills and resources to their communities, articulating particular class and race projects associated with citizenship rights, thereby expanding notions of citizenship in Brazil. This kind of vision was promoted in the activities that took place over events such as the Semana de Consciênçia Negra (Week of Black Consciousness), which took place in November 2005 while I was in Salvador. I attended some of the events taking place in the Vila Velha theatre in the city centre and across various venues in the historical neighbourhood of Pelourinho. One event at the Vila Velha theatre attracted well-to-do African descendants, including students, activists, scholars, and interested members of the public. Vanda Machado, a noted historian and educator, gave the opening address. Looking at the admiring faces in the audience, I thought that these young people and other individuals who she influences were growing up with a strong sense of ethnic identity and pride, and possibly thought of themselves as forming part of an up-and-coming black intelligentsia, some of whom are interested in "giving back" to the Afro-Brazilian community.

This was certainly the orientation of research participants such as Silvio Humberto, founding member of the Steve Biko Institute, who has served on the Salvador, Bahia, city council, being one of the more recent members of the progressive wing of the Black Movement to be elected to local government. He manages an organization that provides after-school educational programmes for black public school students. One of the programmes is called Orguntec: it is partially funded by an African American scholar from the University of Harvard, who, in liaison with Livio, was interested in increasing the number of blacks in medicine, engineering, computer science, math, physics, and the biological sciences, where they could receive better remuneration. Here, there was a deliberate attempt to create a class of educated and successful black people who were knowledgeable about their rights. The Steve Biko Institute has also developed educational materials, such as "Anti-Racism and Human Rights Program for Young People," which aims to give students a firm grounding in their racial and cultural identity, preparing them to be vocal about issues of racism and discrimination. Certainly, when it was formulated in the 1990s, the Steve Biko Institute was a revolutionary project as black students and their parents simply had no real options in Bahia's educational system regarding college admission and affirmative action. Livio and his collaborators had to critically rethink the role of education for historically oppressed and excluded communities, including articulating a more positive valuation of the contribution of Afro-Brazilian culture and people to Brazilian society; deconstructing Brazil's and Salvador's past by explaining slavery and its negative effects; challenging the myth of racial democracy; and denouncing racism within the context of international human rights law (Dixon 2016). In focus group discussions with the students enrolled in the programme, most said that their lives had changed for the better—but the discourses of the Black Movement, which they learned through the programme, also put them into conflict with their families and communities (see Chap. 6). Parents, siblings, and neighbourhood friends would often comment, "*você está ficando maluca*" (you are getting crazy) or, in discussions about specific Afro-Brazilian traditions, some would say, "*Tira essa macumba dai meu filho*" (get rid of this macumba talk, my son).[3]

[3] "Macumba" is a term that has been used to describe African-derived religions, usually in a negative way to mean witchcraft and black magic.

The students expressed a desire to help their families and communities but also had conflicting ideas about staying in the neighbourhood once they achieved mobility. They expressed the view that they would only live in their community in fear, because people are always observing you when you start to achieve things and become envious of their position. The work of the Steve Biko Institute linked pre-vestibular courses to education and affirmative action and fed into larger discussions on racial inequality, citizenship, democracy, and social justice—especially for poor students in Bahia. Along with other similar programmes, it challenged prevailing racial hegemonies in Salvador, especially about how racialized practices worked in higher education.

Another notable organization previously mentioned in Chap. 2 was Didá Educational and Cultural Association, whose Cultural Coordinator is Víviam Caroline de Jesus Queirós. She saw the role of her organization as contributing to preventing the girls from vulnerable communities entering criminality or prostitution:

> For the girls who grew up in this organization, you can see that they are different women. It isn't all the time that you will be victorious. We have seen cases that are sad, that make you cry. Especially where you make an extra effort with that girl, and what happens is not what you wanted for her. However, most girls in the organization think different, feel different, and confront life differently to other girls in the community. They are girls involved in projects and associations and are doing something with their lives. The other girls in the community are just hanging out in the doorways, walking up and down the street. But girls in this organization walk and talk differently, have a different way of expressing themselves. (Interview: June 29, 2017)

Other research participants volunteered their time in social programmes or associations in the community or donated money to Candomblé temples or schools. Jorge Carvalho is an interesting case. He built a successful career in the construction industry before deciding to freelance in the fast-growing and profitable culture industry, which was heavily subsidized by the state at the time that I conducted fieldwork in Salvador (state funding has considerably declined since). The culture industry, linked to tourism, promotes "Bahian cultural heritage" and particular aspects of black culture and folklore such as Capoeira, Afro-Bahian cuisine, music, and Candomblé. Although providing fresh opportunities for social mobility, it appears to reproduce traditional forms of power and authority in that the newer, vibrant Afro-Brazilian cultural organizations tend to get funded and patronized by the white elite. Jorge's connections with both the black leadership in these organizations and his former integration in white elite networks through the construction industry are considerable assets in this line of work. As mentioned previously, Jorge formed an association of black businessmen specifically to address the lack of mentorship for black people in this area—an association that attracted around 50 young people at the time, most of them from disadvantaged backgrounds. Jorge's strategy was to focus on "ethnic niche markets"—markets focused on the packaging of "Afro-Brazilian" arts and crafts, mostly for tourists—as well as cosmetics used by dark-skinned Brazilians. There is real potential for this kind of market, since the state was willing to fund such projects through a redistribution programme aimed at black, poor communities. He claimed that his motivation for starting such an enterprise was his concern for economic democracy for black Brazilians. Indeed, participation in these spheres has provided some opportunities for upward mobility, while conferring social status of a certain kind.

Other research participants highlighted their work in the black community through their work, such as one informant, Minister Luislinda Dias de Valois Santos, Minister of State for Human Rights and former Head of the Special Secretariat of Policies for the Promotion of Racial Equality for the Ministry of Justice and Citizenship. Considered one of the first black judges in Brazil, she was the author of the first verdict condemning racism in Brazil and developed important projects, such as "Justice, School and Citizenship," which focused on young Brazilian students. In 2011, she became a judge of the Court of Justice of Bahia. As head of the Special Secretariat of Policies for the Promotion of Racial Equality of the Ministry of Justice and Citizenship, she initiated the action of conflict mediation between police authorities and vulnerable communities within the scope of the National Security Plan.

LUISLINDA: I am involved in many ways in the black community. Because I came from a poor layer of society, a neighbourhood that is a black community, São Caetano, and I was born with a mission to take care of people, I do this with love and care. I grew up in Capelinha de São Caetano in Liberdade. This is a very strong black cultural area.

INTERVIEWER: Are you personally involved in community projects?

LUISLINDA: Yes. I work a lot with women prisoners, I do a project with my sister in schools, called "Schools and Citizenship." And another project that encourages reading and learning while waiting for justice. Another one was about putting books at bus stops. I chose the books myself. The "Reading, Learning, and Waiting for Justice" programme is a project to provide books as people are waiting to have their claims heard for justice. I also provided chocolate, so people could enjoy their wait! It was meant to encourage people to feel good and get something good in their minds instead of being bored. I had a partnership with Livraria Cultural and I used to buy newspapers and magazines to share with these people. Together with reading, I provided chocolate, to give something sweet while they waited. I did these projects while I was at the Tribunal of Justice in the state of Bahia. I bought everything with my salary. (Interview: June 28, 2017)

However, it can be problematic if one is thought to act against the interests of the black community. Túlio Santana, the director of the transport syndicate in Salvador and a member of the Black Movement, was once described by another black activist as a *vítima do narcisismo branco* (victim of white narcissism). He was accused of being a black man who had adopted white capitalist values, charging higher bus fares that would disproportionately hurt the Afro-Bahian poor. Some considered him to be a traitor to his own race, living the life of a white upper-class person with a comfortable apartment, private education for his children, frequent vacations, marriage to a lighter-skinned professional, and shopping at exclusive malls. Such a critique, regardless of how appropriate it may be, mixes an accusation of class oppression with one of race treason. Ironically, the movement that he seeks to align himself with rejects him. Yet another member of the Black Movement, Vanda Machado, a well-known educator, has a similar lifestyle in white neighbourhoods, yet she is not subject to the same slander that Túlio is. This is probably because Vanda demonstrates deep commitment to some valued public signs of blackness, such as her involvement in Candomblé, which has received sate recognition and support after centuries of repression. Túlio does not consider such signs to be an essential part of one's black identity, describing this kind of positioning as "too radical." Vanda apparently practices the right kind of racial-cultural politics in this black activist milieu since she has learned how to balance her political stance and various public roles.[4]

Attitudes towards black leaders in the community underscored the ambivalent and contradictory relationships between the black middle classes and black working classes. For example, in a conversation with one of the smaller Afro-Brazilian cultural and educational organizations located in Pelourinho, the leaders acknowledged that there are black people in good situations, but that historically, black people were not philanthropists in their communities. They discussed various well-known black personalities, such as the Minister of Culture, Gilberto Gil, and the musician, Carlinhos Brown. While acknowledging Gilberto Gil's work as a politician in valorizing Afro-Bahian culture and heritage through the "Ponto de Cultura," a programme through which Afro-Brazilians could access funding for their community projects, he was not personally seen to have contributed to the community. When I raised the point about Carlinhos Brown setting up the Afrodrome in 2013—a new, alternative space dedicated to the Afro-blocos—their response was:

> Yes. He ended up doing the Afrodrome, but everything was thrown together into one space and one commercial context. And it didn't have the visibility that it needed to have. I felt excluded, because there were only Afro-people. The whole world needed to be there paying homage. Then they put us to model/parade in the commercial space and there wasn't the necessary aid. So, it was complicated. Not that the idea was bad, but I think the government bodies could have done more, taken this idea of Carlinhos Brown and strengthened it. Because the idea of the Afrodrome is relevant. However, we needed to parade/model not only in the commercial space there, but also on the main avenue, at Barra … but they make the Barra route available only to the bourgeois and the Afro-blocos are supposed to stay in this commercial space? Why? Why can't we parade on Barra? Why not? Why must you exclude and isolate? (Interview: June 26, 2017)

[4]Aspects of these arguments were developed in previous publications of the author, though significant recent ethnographic data has been added. Please see: "Negotiating inequality: The Black Middle Classes in Salvador, Brazil," in *People, Money and Power in the Economic Crisis: Perspectives from the Global South*, 106–128, edited by Keith Hart and John Sharp (2014, Berghahn Press).

In this section on community participation, I attempted to reveal the different modes of community building and how some participants become involved in their community, revealing the complex relationships between socially ascendant black people and the working-class communities they continue to be associated with. Religious groups also played a role, as struggles over religion (especially Candomblé, Catholicism, and Protestantism) linked up with debates about the formation of Afro-Brazilian identity in Salvador and reflected historically important avenues of social mobility as well. These struggles have been documented in more detail elsewhere (e.g., Selka 2007; Gordon 2013)—below I will focus on a few examples.

Religion, Social Mobility, and Identity

Historically, religion was identified as an important sphere of social mobility for Afro-Brazilians (De Azevedo 1996; Pierson 1942). The Roman Catholic Church has long enjoyed a large and diffuse influence on Brazilian society, though it has been linked to the elite (Borges 1992; Selka 2007). The Catholicism that developed in Brazil, however, was one that had more in common with folk practices than with the official orthodox doctrine of Rome, leading to popular expressions of Catholicism such as the cult of the saints, communal processions of faith, and specific family rituals (Borges 1992; De Queirós Mattoso 1992). Membership in the Catholic Church may align with the need for some upwardly mobile African descendants to belong to a traditionally dominant, moral community with social acceptability. While the Catholic Church rarely challenges the racial hierarchies and power structures in Salvador (exceptions are the Liberation Theology tradition, Inculturated Mass celebrated at the Rosário Church in Pelourinho every Tuesday evening, and the Pastoral Afro programme, see Gordon 2013, 227–228), there exists a kind of moral principle whereby the poorer members of the Church can legitimately access and manoeuvre the social hierarchy without shame and embarrassment, in contrast to their experiences in other areas of life. Thus, there is a well-established code of ethics in the church in which well-off people have the responsibility for caring for less well-off people.

Furthermore, religious rituals within the church (such as first communion, prayer meetings, and youth groups) put people in touch with communities of support within the religious sphere and conferred respectability and morality on families and individuals. Church teachings and rituals produce particular, moral selves, properly socialized persons who know how to conduct themselves and treat others with *amor* (love) and *respeito* (respect). This was important for black families seeking to distance themselves from negative stereotypes of blackness. For example, Norma Sanchez's membership in the local Catholic Church offered a haven after she went through her divorce. She could protect herself from accusations of being a *puta* (sexually loose). It was primarily women who made claims to respectability through the church.

In the past, Catholic brotherhoods and confraternities helped newly arrived Africans to adjust and assimilate to Brazilian society and represented one of the few organizations through which black people could ascend to leadership positions. It also offered a legitimate way of organizing around African ethnic identities and resisting slavery by arranging to help slaves to buy their freedom as well as offering financial assistance. However, the importance of these associations began to decline after 1870, indicating an increased secularization of life. One of the Catholic brotherhoods that has survived to this day is the Irmandade Nossa Senhora da Soledade Amparo dos Desvalidos (Our Lady of Solitude, Helper of the Destitute), which transformed itself into a secular mutual aid society in the post-abolition period. Today, it functions as the Sociedade Protectora dos Desvalidos (Society for the Protection of the Destitute). Now that the state has taken over many social welfare functions, the organization dedicates itself to education and career training for poor youth who are in the majority black and mixed race. Another surviving institution is the Irmandade Nossa Senhora do Rosário dos Pretos, which is housed in the same church of that name. Now a symbol of black consciousness in the historic city centre, there are often kinship links among the members and the association has received the increased support and attention of African Americans in the United States interested in the organization's continued survival. Figure 5.2 is a photograph taken by the author of the Church of Our Lady of the Rosary of the Blacks in Pelourinho. The church was built by slaves in the eighteenth century to provide a place of worship for blacks.

Nevertheless, these older organizations struggle to survive amidst the emergence of newer organizations. In an interview with a senior member of the Sociedade Protectora dos Desvalidos, he stated that:

> The newer organizations are more cultural and linked with the publicity of carnival, so they are full of pomp and show. Obviously, they attract more

Fig. 5.2 The Church of Our Lady of the Rosary of the Blacks in Pelourinho, Salvador. (Source: Doreen Gordon)

> money for their programs. But we have always had a very strong racial identity and continue to serve the poor in the community. We have expanded our concept of poor people to include black people, mixed people, and whites. These new organizations act as if they are only composed of black people, for the black people. But in the past, as today, our associations can't be successful without the help of whites. I also think that organizations such as Ilê Aiyê reinforce traditional stereotypes of Afro-Bahian identity, such as we are only good at dance, music, performance, art … while not paying enough attention to education and technical training which would really improve the situation of black people. I do not think these newer organizations do much to improve the range of things that black people can do and the different ways that they could try to lift themselves out of poverty. (Interview: May 4, 2006)

This quote reveals the tensions between older and new forms of association and ethnic identification. The newer associations attract a section of the black elite, who define themselves according to the more contemporary formulations of blackness and are generally made up of younger people. The newer, bolder Afro-centric cultural organizations have more of a monopoly on opportunities for social improvement and upward mobility through avenues that have been opened to expanding tourism, music, and culture industries. Thus, the older organizations must seek new strategies of survival and self-definition that will sustain them in the present and future environment.

Perhaps a more obvious connection with the concept of blackness and anti-racist practices would be the Afro-Brazilian religion, Candomblé. On a cool evening in May 2007, I visited one of the oldest temples near Salvador's city centre. The temple of Gantois was founded by Mãe Menininha, one of Brazil's most famous spiritual leaders. When the anthropologist Ruth Landes conducted fieldwork at Gantois in the 1930s, the temple was situated in the woods and was difficult to reach (Landes 1947). Now it is highly visible, sitting on top of a hill overlooking the dense and socially mixed neighbourhood of Federação.

Emerging from repression in the nineteenth and early twentieth centuries, Candomblé appears thoroughly inserted into contemporary Bahian life and enjoys privileged status in Bahia as cultural patrimony. Thus, the influence of the religion goes well beyond the walls of the temple—the signs, symbols, aesthetics, rhythms, and philosophies of Candomblé circulate through various quarters of Salvador's public spaces, increasingly appropriated by the state, Black Movement activists, gay activists, and Salvador's eco-movement. However, not all temples enjoy the same status. Some are well-maintained and funded with spacious properties and resources; others are not. Candomblé's support base remains varied, although the majority of practitioners remain poor, Afro-Bahian women.[5] On this particular evening, several persons alighted from expensive, foreign-made cars in the parking lot provided. The women wore long, flowing skirts and beaded necklaces, many with braided and dreadlocked hairstyles wrapped in African print material. Their perfume left a scent in the air as they walked towards where the singing and drumming was already in full swing. A large crowd was already gathered in the temple. A few women in the typical starched white dress of Candomblé initiates were

[5] A report compiled by Dos Santos (2007) for the Center of African and Oriental Studies at the Federal University of Bahia on the conditions of contemporary Candomblé temples in Salvador reveals that 63.7 per cent of the leadership is female; 88.7 per cent classify themselves as *pretos* (blacks) and *pardos* (browns); and most have low levels of education. http://www.terreiros.ceao.ufba.br/pdf/Os_candombles_no_seculo_XXI.pdf. Accessed July 31, 2021.

dancing in trance, their heads slightly dipping to the left and the right. Later, as I discussed my experience with some Brazilian friends, they smiled knowingly and said that Gantois tends to attract an elite crowd, especially well-to-do black professionals and activists.

Some black middle-class people found in Candomblé a rich source of symbols and cultural referents in their attempt to create a unified black identity based in African traditions. Vanda Machado, a prominent black educator, coordinates a school on the compound of the Ilé Axé Opô Afonjá temple. She designed a curriculum specifically focusing on raising self-esteem and awareness among the children from surrounding communities of their African heritage.

Candomblé temples have been important in the black community because Afro-Bahians have often been excluded or denied land ownership and tenure and forcefully evicted from their lands. Property is often handed down by priests and priestesses to their initiates in training, offering a means of economic security and stability especially for women (Landes 1947; Hita 2014). Candomblé temples not only serve as both permanent and temporary shelters for the leadership hierarchy but also for initiates and persons seeking refuge from poverty, domestic violence, and state agencies, among others. They often redistribute food and services to the community: the more prestigious temples also offer advantageous contacts as well as employment and information on jobs and are significant employers themselves (Matory 2005; Bastide 1978, 171).

However, some research participants viewed Afro-Brazilian religious rituals as backward, superstitious, and barbarous. Despite the recognition that Afro-Brazilian religions receive today, there is still stigma and shame surrounding the practice of these religions. For example, De Santana (2009, 156) notes that his research participants referred to grandmothers, aunts, and uncles who were involved in Afro-Brazilian religions—but not themselves. The choice not to be involved in these traditions may be strategic for upwardly mobile research participants, particularly those in the lower middle classes. This points to the possibility that wealthier Afro-descendants may have more freedom to choose this religion, once their status in the middle classes had been consolidated. For example, Luislinda Dias de Valois Santos, Minister of State for Human Rights, is an open practitioner of Candomblé.

INTERVIEWER: How did you get involved with Candomblé?

LUISLINDA: My aunt Elena, she was a big, strong woman and active in the religion. Her parents were Catholic. Brazil has religious diversity, as expressed in the Constitution.

INTERVIEWER: Did you experience any discrimination as a result of your religious choice?

LUISLINDA: I faced it many times. I have these long necklaces—"*guias*"—that we use in the religion. This is not common for people to use, especially as I represent authority. I'm happy to wear them, as I am happy in my religion. I have no problem in affirming my religion, it is my constitutional right. So, I don't want them to tolerate me but respect me as I respect them. Other religions must be respected too. (Interview: June 28, 2017)

Meanwhile, the membership in Evangelical churches has increased in recent decades, which is a broad term used to refer to a range of Protestant, Evangelical, Pentecostal, and Neo-Pentecostal churches. Indeed, different waves of Protestantism have spread over Brazil since 1832, challenging the dominance of the Catholic Church (Selka 2007, 24; Stoll 1990). Pentecostals in particular are a powerful force, with growing media empires and significant political representation (Birman and Lehmann 1999; Kramer 2005). This has happened even though controversy surrounds their fund-raising efforts, which are often among poor Brazilians. Evangelical churches have achieved remarkable penetration among black and indigenous sections of Latin American populations (Lehmann 1996; Annis 1987). However, it has also appealed to elites, especially as Protestantism was seen to embody the ideals of modernity, rationality, and individualism appropriate to a modernizing country.

Census statistics indicate that many Afro-descendants are members and pastors of Evangelical churches (De Oliveira 2004, 20). This seems to contradict the overwhelming focus on Candomblé as the major religion of Afro-Brazilians and the primary symbol of black identity. This leads us to ask the question—what might be attractive about these religions to socially mobile individuals? Firstly, although Protestant religions are particularly schismatic, they have opened social spaces for new forms of association, authority, and dissent (Stoll 1993, 12). Secondly, they appear to offer a way out of older forms of hierarchy and patronage relationships between white elites and Afro-Brazilians, found in Catholicism and Candomblé. It is possible too that Protestant religions allow members to reduce their economic contributions to community life and keep more of their income for family investment, as Evangelical groups are often critiqued for being anti-social and separating their members from the rest of the society. Through their loudspeakers, their distinctive music, and their open air convocations, they reveal a clear intention to "clean up" local culture (Lehmann 1996).

Protestant Church doctrine tends to condemn Afro-Brazilian culture, putting them into direct conflict with Candomblé practitioners. However, my ethnographic material also revealed instances where they are oriented in different ways towards the Afro-Brazilian community. For example, Baptist churches were heavily involved in the community through the provision of educational services. The setting up of schools as attachments to churches filled the need for parents to properly educate their children, which explains the social mobility of second- and third-generation Baptist families such as Horacio Rocha's family. Horacio was a member of the Progressive Evangelical Movement Movimento Evangélico Progressista—MEP), which consisted of a number of traditional Protestant churches. The elite sections of these churches—lawyers, doctors, public servants, politicians—engaged in a more political critique of social relations in Brazil and have contributed to anti-racist discourse, drawing from Martin Luther King's teachings to advance an anti-racist agenda. Horacio's deep involvement in the Black Movement and issues of social justice were shaped by his religious beliefs. He explains:

> Since I was a child, since my grandmother's time, we had an Evangelical lifestyle, principally in everyday life, in how we lived, our values. This influenced the whole family … such as our ideas about freedom and liberty. We were not fundamentalist. In my family, it was more about how people treated each other and our attitude towards the community. We do not just live in our own personal lives but also in the community, so my family was never closed off from what was happening around us. This was very important to my upbringing. We had a tradition of literacy as well, for reading newspapers, magazines, books … for critiquing the society. So, in our family, faith represented a more diverse cultural experience. My Evangelical upbringing was also important for my relationships with people. We have respect for others because we were taught that it was our behaviour and comportment that defines us as Christians. I was involved in many student

> protests in university. I was a left-wing militant and involved in workers' protests as well. I have always had a more liberal point of view because of my family's engagement with the community, so we were associated with the more progressive Evangelicals in our church. Until today, I am Evangelical because of the promise I made to God to live a Christian life—which I see as a life of values and commitment to the human condition. Principally this is what my religion does for me. (Interview: November 18, 2006)

As exciting as the work of MEP is, however, it has to be considered against the strongly anti-Candomblé thrust of neo-Pentecostal churches such as the Universal Church of the Kingdom of God (Igreja Universal do Reino de Deus, IURD). The leaders and members of this church are definitely not happy about the focus on Candomblé. As part of the IURD's Gospel of Health and Wealth every Friday, the church exorcizes Candomblé spirits, associating them with the devil. There have been numerous incidents reported in local newspapers about Pentecostal attacks on Candomblé temples and shrines, which anger influential portions of the population, including black activists.

Other Evangelical groups appeared to be engaging in a more radical politics of race. There is a movement that began growing from around the 1980s, that includes a diverse section of traditional Protestants as well as members from newer Evangelical churches, who have formed the Black Evangelical Movement (Movimento Negro Evangelico, MNE). They include groups and individuals who have been devoted to issues of race and racism and who have been promoting their activities through the internet and social media across Brazilian cities. Key leaders began devoting concentrated intellectual energy to articulating an anti-racist message in clear, theologically sophisticated terms. This points to the varied and rich ways in which Afro-Brazilians are constructing alternative black identities and the ways in which they are contributing to the anti-racist struggle in Brazil. There is good reason to pay attention to such groups. Firstly, they are gaining new ground—especially among young people—extending their reach and appeal through new information technologies, communications media, and musical styles which incorporate Hip Hop, Samba, and Reggae (Burdick 2005, 2008, 2013). Secondly, they display a radical engagement with blackness. Thirdly, a good number of their leaders and organizers appear to be educated, upwardly mobile African descendants who are articulating a message of black consciousness, Christian responsibility, and socio-economic improvement of the black community which might resonate with a larger number of African descendants than the Black Movement organizations have been able to reach.

A branch of the MNE in Salvador call themselves The National Council for Black Christian Men and Women (Conselho Nacional de Negras e Negros Cristãos, CNNC). In 2006, when I got to know some of the leaders and members of this organization, they had a membership of just over 2000 people, many of whom are educated and had access to technology such as the internet. One of their main concerns was to identify "whitening" processes in their churches and society that demonize African traditions and assign a negative value to the black body and physical features. I attended a few of their meetings, such as in November 2006 at the Colégio Estadual Duque de Caxias, a well-known public school in Liberdade. The stage and auditorium were decked in the colours of the Jamaican Rastafari religion and symbols of black resistance, including those of Afro-Brazilian cultural organizations. Speakers at the event emphasized the teachings of Pan-Africanist thinkers, black civil rights leaders such as Frantz Fanon, Marcus Garvey, and Cheikh Anta Diop and Brazilian black activists such as Abdias do Nascimento. Here, there were clearly contributing to the debate on issues of race, racism, and social justice in church and society.[6]

Discussion

In this chapter, I attempted to illustrate for the reader the varied ways in which research participants experienced class as it intersected with other identities, such as race, gender, and religion. The focus was on lived experiences and the symbolic struggles that participants engaged in as they identified and differentiated themselves from blackness and/or the black working classes associated with poverty. My aim was also to show how several research participants continued to be involved in their communities in meaningful ways, though these efforts were varied, contradictory, and involved critiques from those in lower socio-economic positions. These included involvement in Afro-Brazilian and religious organizations, which also provided selected avenues for social mobility which sometimes reproduced older patterns of patronage between elites (both white and

[6] Aspects of the arguments used in this section were developed in a previous publication by the author. Please see "Religion, 'Race', and Emerging Middle Classes in Salvador, Brazil" (*Canadian Journal of Development Studies/ Revue Canadienne d'etudes du Developpement* 34: 221–225, 23 May, 2013).

black) and the Afro-Brazilian poor. However, this also raises questions about whether revenues from popular black cultural organizations, black businesses, and the commercialization of black culture for touristic and other purposes really go back to poor communities where most African descendants live.

Two distinct ways of building black identities and a sense of community emerge from the ethnographic material: one set appears to be based on celebrating an African cultural heritage and patrimony (in a way that can become exclusionary, racialized, and essentializing) whilst another is based on a more universalizing, future-oriented vision of achieving social equality and combating discrimination through rational practices of the self that will erase the past and allow access to professions with higher incomes.

Discourses about achieving social mobility through self-improvement practices are not new in Brazil. One can discern several ideologies here: for example, there are rights-based approaches grounded in a politics of recompense and restitution, especially because of slavery in Brazil. Another approach may focus on a charitable model in which Christian duty means the advantaged should help the disadvantaged, though this may not entail a critical analysis of the causes of disadvantage. Certainly, the politicization of race in Brazil has led to some unexpected convergencies, such as Evangelicals taking a more positive view of African legacies or embracing radical black politics.

References

Annis, S. 1987. *God and Production in a Guatemalan Town*. Austin: University of Texas Press.

Bastide, R. 1978. *The African Religions of Brazil: Towards a Sociology of the Interpretation of Civilizations*. Baltimore: The Johns Hopkins University Press.

Birman, P., and D. Lehmann. 1999. Religion and the Media in a Battle or Ideological Hegemony: The Universal Church of the Kingdom of God and TV Globo in Brazil. *Bulletin of Latin American Research* 18 (2): 145–164.

Borges, D.E. 1992. *The Family in Bahia, Brazil, 1870–1945*, Electronic Resource. Stanford, California: Stanford University Press.

Burdick, John. 1998. *Blessed Anastacia: Women, Race and Popular Chistianity in Brazil*. New York, London: Routledge.

———. 2005. Why Is the Black Evangelical Movement Growing in Brazil? *Journal of Latin American Studies* 37: 311–332.

———. 2008. The Lost Constituency of Brazil's Black Movements. *Latin American and Caribbean Ethnic Studies* 3 (2): 149–169.

———. 2013. *The Colour of Sound: Race, Religion, and Music in Brazil*. New York University Press.
Butler, J. 1998. *Freedoms Given, Freedoms Won: Afro-Brazilians in Post-Abolition São Paulo and Salvador*. New Brunswick: Rutgers University Press.
Cohen, A.P. 1985. *The Symbolic Construction of Community*. London/New York: Tavistock Publications and Ellis Horwood Limited.
Davila, J. 2006. *Diploma de Brancura: Politico Social e Racial no Brasil, 1917–1945*. Sao Paulo: Editora UNESP.
Da Matta, R. 1991. *Carnivals, Rogues, and Heroes*. Indiana: University of Notre Dame Press.
De Azevedo, T. 1996 [1953]. *As Elites de Cor numa Cidade Brasileira*. Salvador: Editora da Universidade Federal da Bahia.
De Oliveira, M.D. 2004. *A Religião Mais Negra do Brasil*. São Paulo: Editora Mundo Cristão.
De Queirós Mattoso, K. 1992. *Bahia: Século XIX: Uma Província no Imperio*. Rio de Janeiro: Editora Nova Fronteira.
De Santana, I. 2009. À Margem do Centro: Ascensão Social e Processos Identitários Entre Negrosde Alto Escalão: O Caso de Salvador. PhD dissertation. Salvador: Universidade Federal da Bahia.
De Santana Pinho, P. 2004. *Reinvenções da Africa na Bahia*. São Paulo: Editora Annablume.
Dixon, K. 2016. *Afro-Politics and Civil Society in Salvador da Bahia, Brazil*. Gainesville: University Press of Florida.
Dos Santos, J.T. 2007. Os Candomblés da Bahia no Século XXI. Centro de Estudos Afro-Orientais da Universidade Federal da Bahia. http://www.terreiros.ceao.ufba.br/pdf/Os_candombles_no_seculo_XXI.pdf. Accessed 7 Aug 2021.
Douglas, M. 2002. *Purity and Danger: An Analysis of the Concepts of Pollution and Taboo*. London/New York: Routledge.
Gordon, D. 2013. Religion, 'Race' and Emerging Middle Classes in Salvador, Brazil. *Canadian Journal of Development Studies/Revue Canadienne d'Etudes du Développement* 34 (2): 221–235.
Hita, M.G. 2014. *A Casa das Mulheres N'Outro Terreiro: Famílias Matriarcais em Salvador-Bahia*. Salvador: Editora da Universidade Federal da Bahia.
Kramer, E. 2005. Spectacle ad the Staging of Power in Brazilian Pentecostalism. *Latin American Perspectives* 32 (1): 95–120.
Landes, R. 1947. *The City of Women*. Albuquerque: University of New Mexico Press.
Lehmann, D. 1996. *Struggle for The Spirit: Religious Transformation and Popular Culture in Brazil and Latin America*. Oxford: Polity Press.
Matory, L. 2005. *Black Atlantic Religion: Tradition, Transnationalism and Matriarchy in the Afro-Brazilian Candomblé*. Princeton: Princeton University Press.

Mayblin, M. 2010. *Gender, Catholicism, and Morality in Brazil: Virtuous Husbands, Powerful Wives*. New York: Palgrave Macmillan.

Medeiros, M.A. 2018. *Marriage, Divorce, and Distress in Northeast Brazil: Black Women's Perspectives on Love, Respect, and Kinship*. New Brunswick: Rutgers University Press.

Moreno Figueroa, M. 2008. Historically Rooted Transnationalism: Slightedness and the Expression of Racism in Mexican Families. *Journal of Intercultural Studies* 29 (3): 283–297.

Ortner, S. 2003. *New Jersey Dreaming: Capital, Culture, and the Class of '58*. Duke University Press.

Otovo, O.T. 2016. *Progressive Mothers, Better Babies: Race, Health, and the State in Brazil, 1850–1945*. Austin: University of Texas Press.

Pierson, D. 1942. *Negros in Brazil: A Study of Race Contact at Bahia*. Chicago, Illinois: The University of Chicago Press.

Rebhun, L.A. 1999. *The Heart is Unknown Country: Love in the Changing Economy of Northeast Brazil*. Palo Alto, California: Stanford University Press.

Rezende, C.B. 1994. Friendship Among some Young English Men and Women Residents in London, 1991–1992. PhD dissertation, London School of Economics.

———. 2002. *Os Significados da Amizade: Duas Visões de Pessoas e Sociedade*. Rio de Janeiro: Fundação Getúlio Vargas.

Selka, S. 2007. *Religion and the Politics of Ethnic Identity in Bahia, Brazil*. Gainesville: University Press of Florida.

Silva Diniz, A. 1991. O Trabalhador Pobre no Imaginario das Elites Nordestinas (1850–1920). In *Brasil Norte e Nordeste: Estudos em Ciências Sociais*, ed. A.R. Garcia, 49–63. Rio de Janeiro: Editora Brasil America.

Stoler, A. 1992. Sexual Affronts and Racial Frontiers: European Identities and the Cultural Politics of Exclusion in Colonial Southeast Asia. *Comparative Studies in Society and History* 34 (3): 514–551.

Stoll, D. 1990. *Is Latin America Turning Protestant?* Oakland, California: University of California Press.

———. 1993. *Rethinking Protestantism in Latin America*. Temple University Press.

Torres, G.M. 2007. Baiana do Acarajé: A Uniformização do Típico em uma Tradição Culinária Afro-Brasileira. PhD dissertation. Brasilia: Universidade de Brasilia.

Williams, R. 1961. Structure of Feeling and Selective Tradition. In *Raymond Williams on Culture and Society*, ed. J. McGuigan, 27–46. London: Sage.

———. 1977. *Marxism and Literature*. Oxford: Oxford University Press.

CHAPTER 6

It's in the Blood: Tension and Support in Families

Extensive, detailed research on lower-class black communities in the New World have revealed the importance of extended kin networks and the flexibility and adaptability of black families to everyday survival in conditions of marginality (Smith 1956; Gutman 1976; Clarke [1957] 1999; Hita 2014; Queirós Mattoso and Amado 1988). There have been long debates as to whether these adaptive strategies (such as extended families, weak conjugal bonds, close mother-child relationships, and child-shifting) of black families are the result of African retentions, slavery and plantation life, adaption to poverty, or urban marginality (Clarke [1957] 1999); Frazier 1942; Henriques, 1968 [1953]; Herskovits 1941, 1943; Herskovits and Herskovits 1947; Mintz and Price 1976). This has rarely been expanded to include detailed ethnographic studies of how African descendants manage to ascend socially and whether these family patterns disappear as they establish themselves in higher social positions. For example, it would be expected that upward mobility would entail shortening and reducing social networks and kinship solidarity as these can drain resources because mobility often requires channelling them into more concentrated circles of kin. In this chapter, I draw on in-depth ethnographic data to give some insight into the social processes and familial strategies underlying social mobility for research participants.

Upwardly mobile Afro-Brazilians are frequently represented in individualistic ways: as those who have largely "made it" based on their own personal efforts and abilities. Often they are portrayed as leaving behind

D. J. Gordon, *Blackness and Social Mobility in Brazil*,
https://doi.org/10.1007/978-3-030-90765-5_6

their poorer and darker-skinned relations, while marrying advantageously. This popular discourse, present in both everyday talk and academia, tends to result in an abstraction of the upwardly mobile African descendant from their families, communities, and wider social networks in which they live and negotiate their identities.

Pierson (1942) and Azevedo ([1953] 1996) both allude to the familial dynamics underlying social mobility in Salvador in the past. For example, Pierson discusses the rise of the "mixed blood" individual through pathways such as being brought up and educated by white fathers, being the result of common law unions or marriages with lighter-skinned or white persons, or through ties of patronage to the white elite. All these strategies imply processes of "whitening." Particularly fortunate, he noted, were the children of priests. Their fathers, being ordinarily from upper-class families and possessing money and prestige, transmitted to their offspring, coloured as well as white, "advantages of ancestry, social position, intellectuality, and wealth. The professions were readily open to them, and marriages into the bosom of the most exclusive families sometimes took place" (Pierson 1942, 159–160). Similarly, Azevedo describes processes of "whitening" in the integration of mixed-race persons into higher socio-economic positions. In particular, he noted the importance of the protection and help of fictive kin such as godmothers (*madrinhos*) and godfathers (*padrinhos*). He wrote that

> One of the mechanisms that facilitate this integration is the protection and help that many godfathers and godmothers give to their coloured affiliates, educating them … obtaining jobs for them … or assisting with the secondary school and university education and, many times, continuing to orientate them and protect them. (Azevedo [1953] 1996, 166 *my translation*)

However, while Pierson and Azevedo make references to the kinship and personal networks which facilitate social mobility, this is rarely developed into a fuller, more systematic analysis of this area of social life, including the meaning and significance of these networks and how they worked on a practical, day-to-day level.

In this chapter, I focus on the familial context, deploying the word sentiment rather than emotion in discussing family relationships because the first implies not only a feeling but a feeling *about* others. Sentiment also suggests a preference, taste, or predilection for ordering social relations in certain ways. Sentiments, like emotions, are similarly intense and elusive parts of daily life, and both correspond to particular orders of value, meaning, and social relations—but sentiment is more broad-ranging and ongoing. Although family sentiment brings the notion of love to mind, sentiments may be both positive and negative in their content and effects. They are a factor in how people act to create both solidarity and distinction, unity and hierarchy. These factors are certainly at work in the families I interviewed and got to know. While the structures and practices of the family often reflect the society's hierarchical principles, family members also reject and recreate these dominant discourses and practices. In this chapter, I explore how some of these tensions and contradictions play out within the family.

In the following paragraphs, I will show that strategies which are basically shaped for survival in Afro-Brazilian families can be adapted for mobility, that becoming middle class does not necessarily imply pulling up the drawbridge after oneself, and that there remains a strong conflict between directing resources inwards for a smaller family circle and distributing them outwards to a wider one. While this affects the recently ascendant, even relatively well-established black middle-class families must negotiate this tension. While some of these patterns and experiences may not be peculiar to black people, my material shows that race shapes the contours, experiences, and possibilities of social mobility in particular ways.

Families and Class

Historically, the term "*as famílias*" (the families) in the Bahian context was associated with an expanding middle sector of the society that included some African descendants, as distinct from the undifferentiated masses associated with blackness (Borges 1992, 36). Given the high frequency of consensual unions in Bahia (Frazier 1942; Queirós Mattoso 1992; Borges 1992), at least some of these families originated in informal unions, inciting anxieties about race, birth, and legitimacy in the construction of bourgeois selves. Thus, practices of distinction involved stereotyping the lower classes as being the result of "immoral" procreation of the "de-cultured" descendants of slaves with the properly constituted families of the middle classes. However, according to Borges (1992, 52–53), the extended families of the Bahian elite stretched to selectively include illegitimate members of the family as well as dependents of all sorts. This suggests that despite an entrenched system of privilege which favoured a small white elite, social relations across race and class differences created spaces for the social ascension of African descendants, leading to the widespread representation of Bahian social relations as being free of racial tension (Kraay 1998; Azevedo [1953] 1996; Métraux 1951).

Long-standing families in the middle classes could point to other families of similar status and may refer to each other as people of "*sangue bom*" (good blood). Thus, people often positioned others according to their family name and their personal networks. An interesting characteristic among some middle-class families in this study is that they frequently maintained close friendships over generations in which their children and grandchildren could participate. Therefore, privileged networks get reproduced over time. This did not mean, however, that the social structure was static—it allowed for movement between the permeable boundaries of social groups and the absorption of new members.

The Euro-Bahian elite, as mentioned before, were largely invisible to most city residents except in some contexts.[1] Through patterns of preferential cross cousin and parallel cousin marriages, the elite developed well-defined groups of kin with a strong group identity, maintaining property and wealth within the lineage over the generations (Borges 1992). It was the perception of some of my interlocuters, especially those from non-middle-class origins, that certain well-placed Bahian families were able to ensure that their offspring were recruited into similar positions, suggestive of a high degree of social self-recruitment within privileged strata from one generation to the next. More specifically, it is a common perception that lighter-skinned people have tended to occupy more privileged positions in comparison to their darker-skinned relations, though recent research has demonstrated that *pardos* are only marginally better off than *pretos*: the real inequality is said to be located between whites and non-whites (Hasenbalg and Silva 1990, 1991). Pedro Santos, husband of Miranda, expressed the view that when he took the *vestibular* to get into the Medical Faculty, the examiners discriminated against him because he had to enter his residential location on the paper:

> I believe that I passed the exam. In those days you had to put down your surname … and people look at family names. Especially in faculties like Medicine and Law. Fathers send their children and grandchildren there. I

[1] For example, in political life, high-profile weddings, art events, and elite sports such as car racing.

got a lawyer to investigate my case but after a while, I became annoyed and could not be bothered. (Interview: October 8, 2005)

Pedro's view was that there was an entrenched structure of privilege (the reproduction of professions) that acted as a deterrent to black social mobility. Whether or not Pedro's inclination was correct, patron-client relationships in Brazil have tended to mean that well-to-do families whose members knew each other arranged for each other's welfare. By doing so, they kept power, opportunities, and resources within tightly knit, elite families.

However, family sentiment (inclusive of fictive kin and distant kin) can work across lines of difference to aid the social mobility of some individuals. For example, Nubia, an attractive, middle-aged woman who owns two restaurants in the city, related that as a young girl, her mother sent her to live with relatives in Pernambuco after the death of her father. One day she met the director of a private school who could trace kinship links to her mother's family. Based on these lines of common blood, he generously offered Nubia a free place in the school and provided her with books and school supplies, enabling her to eventually train as an elementary school art teacher.

As the city transformed and modernized, the relationship between long-standing families and structures of privilege (such as access to jobs, social networks, property, and wealth) became increasingly destabilized and less straightforward. Access to land became more urgent with increased rural to urban migration and the peripheralization of the city, with resultant family tensions in the distribution of wealth in the form of property or investments transmitted inter-generationally (Hita 2014). Deciphering a person's origin, social background and class status also became more difficult. Changes in social values, higher educational achievement, and the impact of media and information technology (among other factors) create high expectations for job satisfaction, greater opportunity, and a better quality of life, especially among younger people. Sources of status and prestige were clearly altered as rapid urbanization and changes in social values took place. One of the factors that helped to stabilize families through these changes was extended family networks, though these ties could also be a source of inequality and exclusion.

Ties that Bind

Recent studies of kinship patterns in the New World have emphasized the extended, flexible, and adaptive nature of black families and the ties of cooperation and mutual aid among kin members which provided a level of stability in circumstances of marginality (Stack 1974; Gutman 1976; McAdoo 1981; Engram 1982). These scholars were eager to refute older arguments about the dysfunctional nature of black families, which included references to the unstable nature of marriages, high percentage of consensual unions, and illegitimate children, leading to female-headed households mired in poverty. The revisionist work of scholars such as Gutman (1976), Higman (1976, 1978, 1984), and Jemmot (2015) revealed that there was a stable family system going back to slavery and that researchers had exaggerated the preponderance of female-headed households and failed to look at the development history of black families (see Gutman 1976, xviii).[2] Instead, they show how strong this family system has been, despite the immense obstacles which slavery, poverty, and urbanization have put in the way.

Scholars such as Stack (1974) have written about black extended families in the United States as something akin to moral economies based on sharing and claims making. She emphasized flexibility, flux, and elasticity of residence; the expansion and contraction of households; the successive combinations and re-combinations of kinsmen living together (and often across geographic households) as a coping strategy among poor, black people. Individuals who are members of these extended families are said to align themselves in order to provide for the basic functions and needs of members, adapting flexibly to changes and to daily social and economic problems. Thus, instead of being highly unstable, Stack showed that black families cultivated extensive networks of kin and friends who could support and reinforce each other, devising schemes for self-help, strategies of survival in conditions of severe deprivation, and offering each other temporary child care exchanges and support (Stack 1974, 28). Yet at the same time, the inherent reciprocity of the obligation to help kin required that members keep their resources fluid and open to emergencies of those within the extended circle. This kept families stable but presumably did not allow for upward social mobility.

[2] A careful summary of this literature, as well as its different threads, is provided by Christine Barrow (1996).

The argument that extended families helped individuals cope with poverty suggests that these kin patterns cease to exist above the poverty level. Furthermore, some researchers have argued that life in urban areas has hastened the erosion of extended family values. Urban values of individualism, secularism, and materialism have been said to threaten the strength of black kinship ties (Martin and Martin 1978, 89). However, my material shows these family patterns continue to be a viable element within Afro-Brazilian family life even as families move upward and consolidate themselves. Thus, I argue that extended family patterns are more than just a coping mechanism out of poverty. Although social mobility was often attributed to the hard work and sacrifice of individuals themselves, some advancement came through the support of different family members who invested in a family member's mobility.

While strong family sentiment characterized extended family networks, people are selective about their involvement with kin. *Festas de família* (family parties celebrating life cycle events such as baptisms, graduations, weddings, baby showers, and birthdays), annual festivities, and family vacations presented opportunities for regular interaction and for the creation and re-creation of kinship and friendship bonds. Among middle-class families, these events are usually celebrated privately within the home and often spread out between the homes of different relatives. On a more practical level, both ritualistic and everyday interaction allow for a redistribution of information, support, connections, job opportunities, and so on. People spoke about how they learned skills from their kin (e.g. dress making and cooking from their mother or mechanical work from an older brother) from which they were able to earn money or they inherited family businesses (e.g. in one case a son took over the operations of his father's transport business). Exchange of information about work, employing each other in jobs, and passing on knowledge and tools/skills between family members and close family friends are some of the ways in which family help each other out and promote a family member's mobility.

Working and caring for others emerged as central themes as these families sought to better their lives. Labouring or "suffering" (working hard) for others, particularly one's family, draws on powerful cultural norms and ideals about proper sociality and morality, expressed in popular sayings such as *quem não respeita a mãe, não respeita a gente* (who does not respect his mother, does not respect us). Someone who is unprepared to work or care for kin (especially one's mother) suggests that they have no consideration (*consideração*) or respect (*respeito*) for others. This acquires certain racialized connotations as a person becomes upwardly mobile. Such a person should have an awareness (*consciência*) of their position in relation to their relatives; otherwise, they could be labelled as snobbish and as having "whitened." If a person of lower status is seen trying to get the attentions of the upwardly mobile individual, he or she might also be labelled as trying to get above himself, as "trying to be white." It is precisely the need for such internal cooperation within extended families which can generate very intense emotional conflicts. There is tension between wanting to be included in social relations and the need to assert autonomy and independence, presenting particular dilemmas for black people as they risk accusations of abandonment, vanity, and wanting to be white.

Nevertheless, family members have ways of bestowing status on their relative. When someone shows signs of advancing, they are generally looked upon with respect, esteem, and favour for breaking into mainstream "white" society. When someone in the family advances, this creates the hope that other members can also become socially ascendant—they consider themselves a little more successful because the person has made it. Reconciling the issue of helping one's family with the need to avoid appearing too proud and individualistic is not always easy, as many informants desire to be accepted by family members and some never felt at ease in the primarily white neighbourhoods to which they move. They may visit their poorer kin with gifts, sponsor a relative's education, or host family celebrations at their houses. Most middle-class family members tried to maintain their high status in the family by making small but regular donations or by helping family members in non-financial ways. To give is also to increase one's power and prestige within the family.

Explicit ideologies of mutual support in transactions between household members and ties of sentiment mobilized feelings of reciprocity and obligation. Thus, Miranda Santos would often say of her godmother's daughter (who was her neighbour), "*Ela é chatinha mais boazinha*" (she is annoying but very good to me). She explained that despite her bad characteristics—such as her aggressive and confrontational way of relating to people—she understood the value of helping and reciprocating favours. For this reason, Miranda described her as a person with a heart, who shows consideration and respect for others. It could be argued that this emphasis on obligation, reciprocity, and the need to respond to claims on time and resources could be a deterrent to social mobility. Indeed, some informants pointed to their diverse roles in different households (as sons/daughters, wives/husbands, uncles/aunts) and the need to help their parents as reasons why they continued to live in the same neighbourhood of their birth. However, family members also mobilized their resources and supported others to become upwardly mobile. Older siblings, in particular, might feel responsible for younger siblings, as they sometimes acknowledged that it felt good "bringing other siblings along with them." As the eldest daughter of her mother, Nubia recalled sending money home to her mother since she was 15 years old and making money from selling sandwiches and sweets. This money was sent to help with her younger siblings. As a mature woman and successful owner of two restaurants in Salvador, she took in and supported nieces and nephews. Indeed, this was a major source of contention between herself and her second husband, who came from a well-established white Bahian family and expressed values centred on self-help and independence. Nubia, however, felt that she could rest her conscience in the sense of having helped her family—"What else is there to do in life but help your family?" she said.

Family members also pooled together to make significant investments, such as a car or a house, which sometimes contributed to the social mobility or status of the whole family. While I was in Salvador, Rosa, her sister, and her daughter pooled their resources to buy an apartment in the more prestigious area of Canela, moving from their original self-constructed house near to a busy transport centre. Family members pooling together to buy *casas de verão* (summer vacation houses) became more common as middle-class incomes stretched to maintain their lifestyles. As Ana Teixeira commented:

> The middle class today is almost finished, poor things. Everything is expensive. We are not moving anywhere. But there are people in the middle classes who buy these houses from fishermen on the beach and slowly, they reform the houses until it has the appearance of a rich house. Everybody in the family contributes to this so they can all use the house at different times. This happens a lot. (Interview: March 15, 2006)

Summer houses add status and prestige to the family and are important markers of distinction, partly because it is associated with white middle-class lifestyles. In the following paragraphs, I turn my attention to exploring the tensions and conflicts within families.

Tension and Conflict in Families

My first meeting with Tarcis (the step-brother of Pedro Santos) had been arranged for his legal office near to the historical district of Pelourinho. I had been warned that he was a very punctual man, so I got up early and took a taxi heading towards the city centre. Unfortunately, the cab was repeatedly stalled in traffic, so I was late by the time I arrived. His secretary quickly motioned me into Tarcis' office. Dressed in a formal suit and tie, Tarcis looked up from behind a large wooden desk and adjusted some papers. "You are late," he said, and I apologized. The office was elegant, with wooden furniture, paintings, bookshelves, and a large glass window offering a panoramic view of the Upper City and the sparkling ocean beyond. His office, strategically located on an important commercial street, was some distance from his modest home in the working-class neighbourhood of Liberdade.

Tarcis and Pedro were the sons of a respected shoemaker whose business was located on Liberdade's main street. The step-brothers had different mothers and so they had a different upbringing. Tarcis had been lucky to not only get a good public education but also have his academic interests encouraged by his mother and maternal grandmother. He entered the Law Faculty of the Federal University of Bahia, a traditional bastion of the Euro-Brazilian elite, while working full-time and studying at night. He commented that "This is the kind of sacrifice poor, black people have to be willing to make if they want to become successful." He insisted that racism had not hampered his wish to improve himself and that Brazil was a racial democracy. Other family members rejected this view, saying that Tarcis had been excluded from university social circles based on white privilege. I found out later that he was involved in one of the oldest Afro-Brazilian organizations in the city, but he clearly felt that a measure of social segregation was compatible with Brazilian racial democracy.

Tarcis' views sometimes upset members of his family, especially his step-brother Pedro who, after completing secondary education, began working in the construction industry, eventually becoming a lower-level office manager in a well-established firm. The brothers often quarrelled. Pedro accused Tarcis of being full of pride (*orgulho*) and of using his knowledge to put people down. Tarcis would say that Pedro was lazy (*perguiçoso*), wasting time with different women when he should be a hard worker and provider (*trabalhador*) for his family. Their versions of each other's life trajectories were contradictory. Pedro's claim that he had been the victim of racism when he applied for the public university examinations to enter the Faculty of Medicine was dismissed by Tarcis:

> He did not study, that's all! Here in Bahia, black people love to party, to go out with their friends, drink and have lots of women. No other place in Brazil has so many parties and festivals as Bahia. Years later when they are still in the same place, they regret the time they did not spend studying hard to achieve something. At this point, it is ridiculous to blame your situation on racism. (Interview, October 16, 2005)

Tarcis also disliked the fact that Pedro was having an affair with a long-time mistress, a lady half his age. He described this behaviour as "*feio*" (ugly), suggestive of lower-class behaviour not befitting his family whose members were trying to better themselves. Tarcis had married early, to a woman he described as *branca*, a book salesperson. He had three children with her, and although the marriage did not last, he saw his primary purpose as being a provider and supporter of his children. All of them, he pronounced, had completed university, had moved to better neighbourhoods, and were doing well in life. This was an implicit criticism of Pedro's parenting practices, as Pedro's adult sons still lived with their parents, had started fathering children outside of wedlock, and appeared to be moving in the same occupational direction as their father.

At time, Tarcis sought to assist family members—such as when he offered Miranda's niece, Marycelia, a job in his practice as a receptionist. However, Marycelia was always late in the mornings, and according to Tarcis, did not seem to understand the work discipline required. Tarcis was not afraid to fire her from the job. Others regarded this as a joke in the family and told the story with great relish whenever there was an opportunity, because everyone knew that Marycelia liked the idea of dressing up for an office job but was not prepared to work.

This case points to several issues. First, the relationship between Tarcis and other family members shows how race often enters the question of social mobility for black Brazilians and involves practices of distinction as well as negotiating a field of claims concerning moral personhood. Second, Tarcis and Pedro both live in working-class neighbourhoods, stretching the preconceptions about what it is to be middle class. Liberdade, however, is one of the oldest neighbourhoods in Salvador, where cultural organizations such as Ilê Aiyê articulate a strong black identity, and community activism brought many services and amenities not available in similar areas.

Several influential persons chose to live in Liberdade—such as the Director of Ilê Aiyê, Vovó—and many give their kinship and friendship ties in the neighbourhood as a reason why they continue to live there. Tarcis, however, was in the process of moving to the new, more modern district of Iguatemi. Family members said that he was building a house with several rooms and a pool, even as they expressed concern about his health, being that he was carrying out this project in his senior years. Yet Tarcis was careful to maintain his roots in the black community, keeping in touch with old friends and relatives, while holding frequent dinners and parties.[3]

Some interlocuters expressed that tensions resulted from their commitment to follow a disciplined path of education and occupational success. For example, I conducted a focus group with some young people who were public school students enrolled in an after-school educational programme that aimed to prepare them for the university entrance examinations. Some children highlighted tensions they experienced in their families and communities in relation to their rigorous study schedule. The programme leaders rationalized long hours of study as necessary because many of the students were behind in their educational curriculum due to the poor quality of the public schools:

INTERVIEWER	Did your family and friends have any reaction to your enrolment in this program?
FLÁVIO	Well, I had to really ask myself whether I wanted a change in my life or not, whether I wanted to accept some of what the teachers were saying. I am the type of person … I have brothers and a mother and father who never wanted anything to do with studying … so, I could have also said the same. It really took courage to have my own vision for my life.
RAILDA	After I entered Orguntec,[4] I used to walk along my street and friends would call out to me, "Do you

[3] Some of these arguments were developed in an earlier publication by the author. Please see: "Negotiating inequality: The Black Middle Classes in Salvador, Brazil," In *People, Money and Power in the Economic Crisis: Perspectives from the Global South*, 106–128, edited by Keith Hart and John Sharp (2014, Berghahn Press).

[4] Orguntec was the name of the educational programme. It aimed to prepare black children from the public school system for the university entrance examinations, especially in science and technology subjects. In addition to the educational curriculum, they taught African history, Afro-Brazilian culture, and personal development courses. The programme also provided students with some meals and help with transportation and school materials.

> want to go to the beach on Saturday?" or "Let's go to the mall." I would always be saying, "No, I have to study." They would say, "You are going to die from studying so much!" Some people no doubt thought that I was snobby and would say things to me like, "*lá vai essa menina*" (here she goes), or "*cu de ferro*" (ass of iron). If there was something on the television and I have something critical to say about it, even my parents would sometimes say, "Everything about this girl is Orguntec! Shut up!"

The focus group revealed similar stories from the other children. Although many of the parents were grateful for the scholarship offered by the program and their children's educational progress, there was also tension surrounding the effort that their new study regimes required as well as clashes in perspectives and values between different family members. This likely explains why some participants were selective in terms of what kin ties to maintain.

The occupational mobility of research participants often entailed moving away from relatives, becoming less social and more career oriented, and meeting new circles of friends. This geographic mobility away from one's former residence and social networks often symbolically associates that person with whitening strategies. Thus, communication can break down, the person who has become socially ascendant may differ in their views on life, interests, tastes, friendships, and so on. Further compounding the issue is that children born to socially ascendant parents may have different values so that tensions between generations and lifestyle can occur. To compensate for a loss of a sense of family and rootedness, some middle-class black people may get involved in organizations, clubs, civic groups, and religious groups in their former communities, which can give a sense of identity and belonging.

Matrifocality

Many research participants emphasized the role of their mother and other female relatives (sisters, aunts, and grandmothers) in their education, orienting them in ways to deal with racism—including rejection (as in Rosa's challenge to racism experienced by her son in Chap. 2), acceptance, or astute negotiation. The term matrifocality has often been used in the literature on black families in the Americas to refer to a kind of family system where the main tie is not marriage, but the maternal blood tie handed over generations (Gonzalez 1970, 237; Smith 1956). The focus is said to be on the primary relationship between the mother and her children, and this usually referred to female-headed households (Hita 2014). Although the term has been criticized for not revealing much about the historical development of the family and the way in which kin and men in particular participate (see Stack 1974; Chevannes 2002; Anderson 2021), it is useful to think about with reference to my ethnographic material.

Many of my interlocuters spoke about the sacrifices and encouragement of their mothers on their path to improve their lives. For example, Bia praised her mother for holding down different jobs and solely looking after her and her brother when her parents divorced. Additionally, Bia's mother worked hard to source educational opportunities for them—and they were able to optimize on those opportunities. Similarly, Jorge Carvalho said that it was his mother who put aside her earnings from her small trade as a dressmaker to send him to Salvador for further schooling, and it was she who encouraged him to further his educational pursuits (he completed a first degree in Economics in the evening while he worked during the day, and later studied for a Master's in Business Management and Economics). In addition to the support and encouragement of his mother, Rodrigo emphasized the role played by his older sisters in pooling their resources to send him to private school where the education was better and in playing a significant role in his educational success, as his sisters would spend time with him going over his lessons and homework. Women also played roles in the inheritance of property—research on female-centred or matrifocal families connected to Candomblé houses, for example, has shown that significant property is handed down between women, which provides a basis for social security (Frazier 1942; Landes 1947; Hita 2014).

A theme in the literature on the black family is the marginalization of the black male and his abandonment of family responsibilities, leading to distrust in male-female relationships (Hita 2014; Smith 1956; Clarke 1999). A constant fear was that of men leaving black women, as formal marriage was associated with middle- and upper-class white Brazilian women and with white respectability. For marriage, black women in Brazil are "the last ones chosen, and the first ones abandoned" in the "awful calculus of the heterosexual romantic economy" (Burdick 1998, 52). At a dinner conversation I had one night with Bia, her mother and brother, Bia expressed her anxieties about her new husband living in Rio de Janeiro while she lived in Salvador. She said that men were *safado* (not to be trusted), to which her brother retorted that women can also be deceitful and dishonest. Danilo said that women, for example, could trap a man into fathering a child that is not his. Fathers who abandoned families, however, were culturally accepted even as this was not regarded as ideal. When Ana Teixeira was younger, she ran away with an older man with whom she lived and had a child. Even though his paternity was established, he infrequently sent financial support for the child once the sexual relationship ended. Ana did not pursue this legally and tried not to make this affect the relationship between the father and their child. Rather, she spoke of how her daughter adored her father. Ana preferred to depend on her own resources and that of her family, only asking the father for money when there was an emergency.

The focus on men's irresponsibility in studies of the black family is said to have overshadowed the important roles that men have played in their families, not only in their roles as partners/husbands but also in their roles as uncles, cousins, brothers, and grandfathers (Anderson 2021; Stack 1974). In DJ Funk Soul's case, his uncle and grandfather played important roles in stabilizing the household finances for his mother who was a single parent. His uncle also paid a large part of DJ Funk Soul's university education and bought expensive equipment for his nephew's music career. Furthermore, in some cases, women actively sought out their children's father's kin especially if this could aid in their social mobility. For example, when Miranda's cousin, Marycelia, divorced from her first husband, she sent her daughter Marta and her son Raúl to their father's mother so that they could be brought up in a higher status neighbourhood and given more access to opportunity.

Child Circulation

Child circulation was also a strategy that I encountered among solid middle-class families, especially as extended family members went through economic hardship, divorce, or spousal death. Often, a child is sent to live with a better-off family member so that they can access more opportunity, such as better education. Extended family relations may be flexible and expandable to include children from previous or outside relationships, formally adopted children as well as *crianças de criação* (other people's or relatives' children brought up in the home). Children born of a previous relationship are sometimes taken into the legitimate household, where they may be accepted without serious question by the legal wife—as appeared to be the case with Rosa's adopted daughter, who family members claim was a product of an affair their father had with a woman. However, the child was brought up in their home and treated equally along with the others. However, when it is the woman who gets married for a second time, sometimes the children from a previous relationship are sent to live with relatives, or they go of their own volition, as happened with DJ Funk Soul. He recalled terrible arguments with his mother's second husband when he was a teenager. Following a particularly bad argument, he decided to live with his great aunt in Liberdade, even though she was in a poorer neighbourhood.

Middle-class families in the generation of Helena's mother regularly brought up other peoples' children: it was considered a good-hearted thing to do, especially if you had the economic conditions to do it. Indeed, this was one of the more traditional ways of negotiating access to opportunity and social mobility, although it was not necessarily seen in this purely rationalistic light. For poorer, non-white Bahians in particular, it could offer a child legitimacy and recognition by incorporating them into respectable middle-class family with a name, effectively whitening the child.

As a school director and teacher, Helena's mother was regarded as a respectable member of her community and brought up a few children in her household, most of them unrelated by blood. They were treated equally along with her biological children and received a portion of her inheritance. Formal adoption was also not uncommon, though it was more prevalent among older generations. For example, Túlio Santana's father was adopted and brought up in the house of a prominent judge who had other adopted children.

Other stories of adoption revealed inequalities and exclusion within the family. When Nubia's father died, she was sent to live with a better-off aunt who promptly put her to work within the house taking care of her children. Nubia cooked, cleaned, and cared for the children, and she was not entitled to the same quality of education or treatment. Nubia explained that her skin colour and poverty likely played a role in this, as her mother, who she referred to as *branca*, went against family wishes and married a black man, casting shame and embarrassment on her family. Nubia was particularly sensitive to the unequal treatment she received and was not willing to remain in domestic servitude. She therefore set about making her own money selling sandwiches and sweets which she used to send herself to school.

Bringing up other peoples' children may be more challenging now, with cramped, urban accommodation, the higher cost of living, modern lifestyles (more money spent on consumption goods and leisure), and the reduction in the number of children younger generations are having. This suggests that this traditional pattern of social mobility is encountering resistance in the current economic environment, which may also indicate changes in social values having to do with the meaning of having and bringing up children.

Fictive Kin and Godparenthood

Godparenthood is possibly one of the best-documented examples of fictive kinship in Latin America (Mintz and Wolf 1950; Pierson 1954; Bloch and Guggenheim 1981; Foster 1967; Kemper 1982; Gudeman and Schwartz 1984). It is an elaboration of the Catholic concept of baptismal sponsorship, as well as sponsorship at confirmations and marriages. In relation to social mobility, it is important to note that some researchers have focused on godparenthood as a way of linking non-related families and a tool for expanding and intensifying social networks (Mintz and Wolf 1950, 348). Individuals might seek to establish ties with wealthier families, thereby opening the possibility of sponsorship and assisted social mobility for the child (Foster 1967; Kemper 1982). Gudeman and Schwartz show that in colonial Brazil, some individuals used godparenthood to secure a possible ally or protector, but that godparents were usually of the same status or a little higher than their godchildren (1984, 45). Interestingly, they noted that colour was historically a factor in the selection of godparents, with almost 70 per cent sponsoring black children (1984, 46–47). Godparent ties may therefore have represented actual ties of support, assistance, and sponsorship for African descendants or the hopes and desires of their parents for a better life for their children.

In the context of my fieldwork, I did not notice references to the gender, colour, or class of godparents. Most tended to be from similar backgrounds and the meaning of these relationships tended to be stronger among migrants from the interior, in strongly Catholic families, and in lower middle-class families. In the case of some research participants, the godparent relationship was only meaningful as a ritual of birth and bringing the child into a community. It was not expected that a relationship of closeness or financial benefit would develop between the godparent and godchild. In other cases, the relationship was close, and some godparents provided substantial help and assistance. For example, in Bia's case, her godmother (a close friend of her mother) offered Norma and her children a place to live when Norma's marriage broke down. They lived with the godmother until the divorce was finalized and Norma received property with which she could establish herself.

Marriage and Divorce

For older generations of women, particularly non-middle-class migrants to the city, establishing their status, reputation, and roots in the city through marriage, constructing a home, and creating a family were particularly meaningful experiences which carried special connotations for black women because of the association of marriage with social mobility and prestige. By her own admission, Nubia's marriages helped her a lot in life. Being an extremely beautiful young woman, she had always attracted men. At 18 years old, Nubia was teaching art classes in a nearby school when her aunt suddenly died. Subsequently, Nubia moved to Salvador and met and married an older black man whom she said helped her to pursue a degree in culinary arts. He was an Evangelical, so she became one too. Her quality of life improved dramatically with marriage, and she had her only child with him—in addition to caring for four children which he fathered in a previous marriage.

Marriage, being a housewife, and looking after one's children were especially meaningful experiences for Rodrigo's mother. She explained that she took pride in keeping a clean house, in making sure that the clothes were clean and pressed, and in cooking the family's daily meals. However, she warned that getting married can be dangerous, because marriage entailed risk, the possibility of physical violence, and unseen events such as long-term illness of one's spouse. She subscribed to the view that it was good to *criar uma família* (create a family)—but that it could also include suffering and so was not to be entered into lightly. For example, she described it as a heavy burden taking care of her husband through a terminal illness.

Both men and women emphasized the need for a good marriage, which they perceived as a partnership between women and men who had equal stakes in it. Maria Humberto underscored that a woman should not take up with *qualquer homem* (any man), like so many women in her neighbourhood did. Here she was re-instating the distinction between herself/her family and the sexual practices of working-class women in her neighbourhood. She stated that a woman needed a partner and companion who she could share her life with and reason with, a man who was responsible towards his family. Her husband, she said, was a responsible man who provided for his family and she had no problems with him. She said that her marriage was *tranquilo* (peaceful).

On the other hand, Norma Sanchez described her marriage as a bad choice, even though her economic situation improved after her marriage. She described her ex-husband as financially irresponsible—he liked to spend rather than save—and that he did not "suffer" for his family (i.e. provide for them). Power imbalances between them meant she had little or no access to his financial resources. However, she began to work part-time and saved that money towards the children's education. She described the breakdown of her marriage as follows:

> I liked him a lot at first. And he liked me. We were lovers for six years. But when he got married to me and I moved into his house, it was a different story. Men are like that. If it was today, I would have thought twice. But we have to experiment, take a risk, try something—that is the only way we will learn and have experience. He inherited four houses from his father—small houses attached to each other on one street. When we divorced, he did not want to give me any of them, he was mean like that. Even his father advised him to give me some of the property. But he fought it, calling me a *filha de puta* (daughter of a whore). You know, when you are separated and no longer married, you are suddenly a whore. These days I do not have any more anger for him. But sometimes I cry, I feel deceived. He was the first man I knew, when I was only 17 years old. The court decided in my favor and when I sold the two houses given to me, I used that money to buy this house. (Interview: June 24, 2006)

Her divorce settlement provided Norma with an economic base with which to take care of her children. Her ex-husband did not pay child support and so Norma had to work. In these struggles to raise her family, Norma was assisted mostly by her kin and her close friends, especially her brother and the godmother of her children (whom she met as a member of the prayer group at the local Catholic church).

Women often experienced downward social mobility immediately after divorce. When Helena Teixeira divorced from her husband, she moved from a large family house to a small apartment, albeit in a prestigious area. They could no longer afford domestic servants and she had never learned to cook—so meals were bought every day from the restaurant across the road. However, her divorce had been more amicable than most: her ex-husband continued to pay her health insurance and left a vacation house for them. For his three daughters, he showed his care for them despite starting a young family of his own—for example, he invited his daughters to family dinners, bought a car to be shared between them, left property for them, and paid for a trip to Europe to spend time with his daughters.

Because marriage and intimate relationships entail potential risks and dangers, ideally women should also guard against the possibility of a failed relationship and being left alone. For example, Bia refused to move to Rio de Janeiro where her husband lived because she would be more dependent on him financially. In Salvador, she had a secure job, so she planned to stay there until her master's degree was completed. She asserted that women must never throw everything away for a man no matter how much in love you were with him.

Women utilized a variety of strategies to ensure the stability of their marriages and families and to promote their children's wellbeing and status. Di Leonardo (1987) describes the "work of kinship" as the set of activities that women carry out to keep a network of family together. This involves "the conception, maintenance, and ritual celebrations of cross-household ties, including visits, letters, telephone calls, presents, and cards to kin; the organization of holiday gatherings; the creation and maintenance of quasi-kin relations; decisions to neglect or intensify particular ties; the mental work of reflection about all these activities …" (1987, 442–443). These activities are part of the "work of kinship" which assists in the fluidity of kin contacts across and within households. They take time, intention, and skill (Di Leonardo 1987, 443). Keeping family together was often important and meaningful for women because of the status connotations of marriage and family and what this meant for consolidating middle-class status. The women in many of the families I interviewed were central figures in this intimate sphere and they were crucial for exchanging information between different branches of the family and in decisions involving the family and household. Female relatives and close friends often helped each other plan and organize children's parties and family dinners, pooling money and resources together with some monetary help from men with the larger purchases.

The discovery of a husband's extra-marital affair could therefore be devastating. Miranda related how she found out about her husband's 13-year affair with a younger woman. One day she trailed him to the woman's house, and on another day, she visited the woman to explain that she was the wife. Her husband had told his mistress that his wife was dead, and maintained a separate household with this woman, helping her to build her house. Overcome with emotion whenever she related the story, she was angry that the other woman's household was prospering while hers was suffering. However, there tended to be greater acceptance of infidelity among older women than among younger women who were more open to divorce, being a single mother, or living with a partner (which after two years constitutes marriage according to Brazilian law). On a return trip to Brazil in 2013, I learned that Bia and her husband had divorced because of an affair he had with a woman in Rio de Janeiro that produced a child. Shortly after purchasing her first home, Bia met another man and recently had a child with him. She sent me pictures by email of the baby's baptism with herself and her partner in the local Catholic church.

Interracial Marriages and Relationships

During my first few months in Salvador, I was invited to an evening wedding to be held in a chapel on an army base in Amarelina. Excited at the prospect of seeing a wedding, I asked my Brazilian friends what I should wear. I was told that it was common to wear your hair high up on your head, with earrings and a nice dress or skirt, with high heels. Luckily, I had a fancy skirt, blouse, and heels. I went to the hairdresser to get my hair intricately put into place. When my friends came to pick me up later that evening, they exclaimed that I looked perfect for attending the wedding. We arrived at a small, white-washed, and beautifully decorated chapel and I was briefly introduced to the parents of the groom. Their eldest son, a dark-skinned man of about 34 years old, was getting married to his girlfriend, whom he had met at university while studying for his degree in engineering. Most of the people in the church were dark or brown-skinned and wore elegant, long dresses with perfectly styled, long, wavy hair. Perfumed bodies bent forward to kiss old friends and relatives, greetings were exchanged, and the audience beamed at the groom-to-be as he entered the church. He flashed a smile as his female relatives smoothed down his black, satin-lined suit and adjusted his tie. An attractive, dignified lady who was standing by the church door was introduced to me as the groom's aunt.

The guests settled down in their seats in the chapel, then stood up to watch the procession to the top of the aisle of *padrinhos* and *madrinhas* of the groom and bride. The women were dressed in bright satin materials that were elegant yet showed off their curves to advantage. The men were dressed in dark suits and bow ties with white or pastel-coloured shirts; they all looked freshly scrubbed and shaved, had their hair cut low, and had applied cologne liberally. They ranged in colour from dark-skinned to light brown. There were a few white faces in the audience who were relatives of the bride. The flower girl and ring-bearing boy then entered, smartly dressed, and trying hard to remember their steps. The groom waited at the front along with his mother, and the congregation sighed approvingly when the radiant bride showed up at the front steps of the church. She was beautifully and conservatively outfitted in a long, white satin dress overlaid with lace and tiny pearls which glinted in the candlelight of the church; she also had on a white veil.

The bride would be described as *branca* (white) in Salvador, as well as her father who walked proudly beside her towards the front of the church to join the groom. The priest proceeded with the wedding ceremony, vows were said, and rings were exchanged, and the *padrinhos* and *madrinhas* signed the witness book. We all held our hands out towards the newly married couple in blessing, and they proceeded to exit the church with the audience throwing rice for good luck.

Such interracial marriages were not uncommon in Salvador and among my group of informants, it occurred frequently enough to seem like a pattern. Indeed, movement to a white neighbourhood and marriage to a white spouse is likely to associate a black person with social mobility and having "whitened" in the eyes of other people. However, marriage statistics suggests that interracial marriages represent a relatively small percentage of partnerships which generally tend to be between people of similar race and class background (Telles 2004; Telles and Esteve 2019; Schwartzman 2007). For example, racially endogamous couples accounted for 68.3 per cent of all couples in Brazil according to 2000 census data (Telles and Esteve 2019, 303). Researchers have shown that Brazilian marriages between blacks and whites are less common, but they occur more frequently than is the case in the United States. For example, in a comparative study of interracial marriage in the Americas, Telles and Esteve (2019, 293) use 2000 census data to show that black–white marriage is 105 times as likely in Brazil as in the United States.

Brown or mixed-race Brazilians out-marry more often than either whites or blacks, which seems to suggest a relationship with "whitening" (Telles 2004; Telles and Esteve 2019). Specifically, brown or mixed-race

Brazilians are four times as likely to marry whites than blacks (Telles and Esteve 2019, 293). According to survey-based research, non-whites tend to marry whites more often when non-whites have higher socio-economic status (Schwartzman 2007, 944; Telles 2004). However, an interesting finding is that higher educational levels do not necessarily lead to an increase in intermarriage as we may expect (if it is our assumption that education leads to more individualistic and universalistic values). Rather, intermarriage between whites and blacks is more likely to occur among men and women with lower educational achievements (Telles and Esteve 2019, 308). This is perhaps shaped by high levels of inequality in Brazil, such that intermarriage is likely to have been especially common in lower social strata, where persons of various hues and phenotypes have coexisted in large numbers and have had greater opportunities to interact (Telles 2004).

Among my informants, the subject of interracial marriage and intimate relationships that may previously have been unquestioned has now become highly problematized because of changed racial politics in Brazil. For example, I was with a group of Brazilian friends at a bar one evening when they began commenting on the personal life and politics of a black gay female academic known to them. They pointed out that this person was once involved with a series of white women—but then as she adopted a more radical black feminist identity, she chose to have intimate relationships with black women instead. They clearly felt that a person's politics, which can transform over time, should not dictate choice of partners. However, if a socially ascendant black man or woman marries a white person, then people also critique this as "marrying up" and "trying to be white." Therefore, when I broached this topic with research participants (many of whom had lighter-skinned partners), they were already on the defensive. They often gave different meanings to the "whitening" narrative that is usually ascribed to such relationships, denying any deliberate choice and claiming that their friends and girlfriends/boyfriends came from the same social circles in which they socialized in school and university. An older informant, the director of a long-standing, historically important Afro-Brazilian association, married a white woman and related to me:

> Well, after you have educated yourself and moved upwards, why would you go back to your community where the majority of women are uneducated? You wouldn't have anything in common with them anymore. So naturally,

> you will choose an educated woman within your social circle, then you can have a conversation. It just so happens that white women are the only ones you encounter as you move up. (Personal Interview, May 4th, 2006)

Túlio Santana met his wife—whom he described as *morena clara* (light-skinned)—in university, explaining that he was attracted to her because of her zest for life—not because of her skin colour. Other research participants joked about their interracial relationships. Rodrigo was dating a Spanish woman when I first met him. He explained that one day, his girlfriend was taking a *concurso* at a university (i.e. a public exam for a job position). She had to prepare a topic on race in Brazil and called Rodrigo on the phone to ask whether he could bring a couple of books for her. Arriving in a hurry to the test-site, casually dressed in shorts and a pair of slippers and sporting a dreadlocked hairstyle, he was suddenly aware of how ironic and politically incorrect the whole scenario might seem, in terms of being the subservient, black boyfriend carrying books for his white girlfriend who was about to take a test on race in Brazil.

According to Rodrigo, the first *branca* (white woman) to marry into their family was his brother's wife—together they shared a daughter, Milana, who was light-skinned. Then his eldest sister, Jussara, got married to a German man. She married at a mature age, when she had all but given up that she would find a husband. When Jussara and her husband decided to visit Brazil, I was invited over for a special family dinner. The matriarch of the family jokingly asked Rodrigo's girlfriend and Jussara's husband to "brighten" the family picture. However even before she got married to her husband and moved to Germany, Jussara had had relationships with Europeans. Her frequent travels as an air hostess took her to countries such as Italy, and she started a relationship with an Italian whom she would visit during the summer and at Christmas. She explained these relationships as the result of the networks she started to move in as she experienced occupational success and became upwardly mobile.

Helena Teixeira described her ex-husband as *branco*. He was a lawyer who worked for many years with Petrobras. Although the marriage did not work out, she claims that he takes very good care of her and their three daughters. When they were married, she met many white upper middle-class friends through him—but also, she inherited the friendship networks of her parents who had white middle-class friends. She also has a boyfriend from Porto Alegre, who calls her his "*negona*" (little black woman). She explains that since these are the social circles she knows, she saw it as natural that she forged relationships with white men.

Interracial marriage between black men and white women was more frequent than the other way around—which is suggestive of gendered valuations of desire. An old Brazilian saying states that white women are for marriage, mulatto women are for sex, and black women are for work. Historically, black women are not considered desirable marriage partners in Brazil (Burdick 1998; Medeiros 2018). In *Blessed Anastacia*, John Burdick (1998) writes about how racism works primarily through the perception of colour and its gradations, producing especially negative consequences for dark-skinned women. Burdick explains the logic whereby women with "bad" (nappy, kinky, and very curly) hair are considered ugly while lighter-skinned women (*mulatas, morenas,* and *mestiças*) are considered sexually appealing. This can produce painful experiences in the spheres of love, dating, and marriage, which Burdick refers to as "the everyday wounds of color."[5] This phrase accurately captures the relentless personal torment that black women experience as a result of being judged according to anti-black aesthetic standards (see Chap. 7).

Inequality Within the Family

Janet Carsten's work (2000) on kinship oriented anthropologists towards relations of caring and sharing in the constitution of kin relations, rather than the fixed, biological, or genetic conceptions of relatedness. However, in some ways kinship in Salvador can be fixed and this can affect access to family resources, the distribution of power among kin, and the socio-economic position of a person. Sometimes, a child who was the product of a previous relationship remains somewhat distant from the centre of the family and may be in a more precarious economic situation, with little or no claim to family property or inheritance. Similarly, in some cases adopted children who do not have *sangue propria* (blood relatedness) had little or no claim to family property. Marycelia, the niece of Miranda, was adopted into the family when her own mother died in childbirth. She is darker-skinned compared to other family members and lives in the same neighbourhood as Miranda. However, despite being brought up by Miranda's sibling who was childless, when she died, she had no claims to the house.

[5] See John Burdick, *Blessed Anastácia: Women, Race, and Popular Christianity in Brazil* (New York and London: Routledge, 1998), especially Chap. 1.

Instead, the house was inherited by Miranda. I asked Marycelia about this and she said it was because she did not have *sangue propria* (however, I wondered whether Marycelia's skin colour also played a role in her exclusion). Consequently, she had little property to bring to a marriage. In Marycelia's case, she often experienced long periods of financial crisis while married and did not have any resources or investments outside of her marriage that could provide her with some leverage. On the other hand, Miranda, as the eldest child in her family, had access to both property and furniture which she brought to her marriage and which also provided her with a measure of independence from her husband. On learning about her husband's mistress, she contemplated leaving him and selling her property to buy a small apartment for herself in Salvador.

It is within the family—that most intimate and inescapable realm of social life, sentiment, and emotional ties—that racial discourses can have powerful effect, as one's physical appearance is interpreted and classified by those closest to you. Racial socialization examines how Afro-Brazilian families engage with practices that simultaneously resist and reproduce racial inequality. In both romantic and familial relationships, love and affection can be distributed in accordance with one's race or physical appearance—or what Hordge-Freeman (2015, 5) has described as the differential experiences of love and affection that reinforce the social relations that contribute to sustaining the racialized hierarchy. For example, Marycelia's first marriage was to a man she described as *branco* (white), with whom she had a daughter; however, the marriage did not last. Today she is remarried to a man she described as *negro* (black). However, at informal family gatherings, she rarely referred to her sons resulting from the second marriage. Maria talked mostly about her daughter—who is lighter-skinned with blonde hair—who was moving up the social ladder as an insurance saleswoman. She indicated that her daughter was beautiful and would "*fazer successo*" (make a success of herself). Colour preferences in families can affect the distribution of love, affection, resources, and opportunities in sibling and other relationships. Here, she makes a link between her daughter's socially acceptable physical appearance, and her potentially successful mobility in the future, underscoring the gendered character of upward mobility in Brazil.

However, respondents had varied responses to racialized hierarchies and discourses. Helena Teixeira recounted that because of experiences of racism that family members had been through in the past as residents of well-to-do neighbourhoods, she worked hard on her children's self-esteem and self-worth. She consistently told them that they were from a respectable and educated family and that they were to ignore racist remarks. They were never to go anywhere uninvited, standing in a corner embarrassed to show their faces. They were to pay scrupulous attention to their appearance, always leaving the house properly dressed in clean and fashionable clothes. Indeed, discrimination sometimes came from their own kind. For example, Helena's brother Bruno remarked that often, he was the only dark face at a party but did not feel racism from the guests. Rather, domestic helpers and butlers often looked down their noses at him. This is indicative of the hegemonic power of ideas about whiteness and wealth that are sometimes accepted as self-evident truths among blacks—they are contestable to some degree and in some circumstances and situations, but they can also be employed among blacks to discriminate, ridicule, and exclude other blacks.

In summary, one of the main points of this chapter was to show how extended family patterns attributed to lower-class black families continued to be a salient feature of family life long after the family consolidated their middle-class status. At the same time, it was clear that family disruption—such as parental illness, death, or divorce—can radically change a family's social status. Higher class positioning could protect in some ways against family disruption—for example, the possession of health insurance and having more money can create more family stability. Extended kinship networks, which could stretch to include fictive kinship ties, could be mobilized to achieve social mobility. Yet this mobility also created tensions that had racialized connotations. In the context of the families I got to know in Salvador, some interactions between kin could be interpreted as claims making which constantly tested, defined, and shaped even close kin relations.

References

Anderson, P. 2021. *Masculinity and Fathering in Jamaica*. Kingston: University of the West Indies Press.

Barrow, C. 1996. *Family in the Caribbean: Themes and Perspectives*. Kingston: Ian Randle.

Bloch, M., and S. Guggenheim. 1981. Compadrazgo, Baptism and the Symbolism of the Second Birth. *Man* 16 (3): 376–386.

Borges, D. 1992. *The Family in Bahia, Brazil: 1870–1945*. Stanford: Stanford University Press.

Burdick, J. 1998. *Blessed Anastacia: Women, Race, and Popular Christianity in Brazil*. New York: Routledge.

Carsten, J., ed. 2000. *Cultures of Relatedness: New Approaches to the Study of Kinship*. New York: Cambridge University Press.

Chevannes, Barry. 2002. Learning to be a Man: Culture, Socialization, and Gender Identity in Five Caribbean Communities. Kingston: University of the West Indies Press.

Clarke, E. 1999 [1957]. *My Mother Who Fathered Me*. London: George Allen and Unwin. Reprint, with a foreword by Rex Nettleford. Kingston: University of the West Indies Press.

De Azevedo, T. 1996. *As Elites de Cor numa Cidade Brasileira*. Salvador: Editoria da Universidade Federal da Bahia.

De Queirós Mattoso, K. 1992. *Bahia Século XIX—Uma Provincia No Imperio*. Rio de Janeiro: Editora Nova Fronteira.

De Queirós Mattoso, K., and J. Amado. 1988. *Família e Sociedade na Bahia do Século XIX*. São Paulo: Corrupio.

Di Leonardo, M. 1987. The Female World of Cards and Holidays. Women, Families, and the Work of Kinship. *Signs: Journal of Women in Culture and Society* 12 (3): 440–453.

Engram, Eleanor. 1982. *Science, Myth, Reality: The Black Family in One-Half Century of Research*. Westport: Greenwood Press.

Foster, G.M. 1967. *Tzintzuntzan: Mexican Peasants in a Changing World*. Boston: Little, Brown.

Frazier, E.F. 1942. The Negro Family in Bahia, Brazil. *American Sociological Review* 7 (4): 465–478.

Gonzalez, N.L. 1970. *Toward a Definition of Matrifocality. In Afro-American Anthropology: Contemporary Perspectives*. New York: The Free Press.

Gudeman, S., and S.B. Schwartz. 1984. Cleansing Original Sin: Godparenthood and Baptism of Slaves in 18th-Century Bahia. In *Kinship Ideology and Practice*, ed. R.T. Smith. Chapel Hill: University of North Carolina Press.

Gutman, H.G. 1976. *The Black Man in Slavery and Freedom, 1750–1925*. New York: Vintage Books.

Hasenbalg, Carlos and Nelson do Valle Silva. 1990. Raça e Oportunidades Educacionais no Brasil. *Estudos Afro-Asiáticos* 18: 73–91.

———. 1991. Raça e Oportunidades Educacionais no Brasil. In *Desigualdade Racial no Brasil Contemporâneo*, ed. P. Lovell. Belo Horizonte: Universidade de Minas Gerais.

Henriques, F. 1968 (1953). *Family and Colour in Jamaica*. London: MacGibbon and Kee Ltd.

Herskovits, M. 1941. *The Myth of the Negro Past*. Boston: Beacon Press.

———. 1943. The Negro in Bahia: A Problem of Method. *American Sociological Review* 8 (4): 394–404.

Herskovits, M., and Frances Herskovits. 1947. *Trinidad Village*. New York: Knopf.

Higman, B. 1976. *Slave Population and Economy in Jamaica, 1807–1834*. Cambridge: Cambridge University Press.

———. 1978. African and Creole Slave Patterns in Trinidad. *Journal of Family History* 3: 163–180.

———. 1984. Terms for Kin in the British West Indian Slave Community: Differing Perceptions of Masters and Slaves. In *Kinship, Ideology and Practice in Latin America*. University of North Carolina Press.

Hita, M.G. 2014. *A Casa das Mulheres N'Outro Terreiro: Famílias Matriarcais em Salvador, Bahia*. Salvador: EdufbA.

Hordge-Freeman, E. 2015. *The Color of Love: Racial Features, Stigma and Socialization in Black Brazilian Families*. Austin: The University of Texas Press.

Jemmot, J. 2015. *Ties that Bind: The Black Family in Post-Slavery Jamaica, 1834–1882*. Kingston: The University of the West Indies Press.

Kemper, R.V. 1982. The Compadrazgo in Urban Mexico. *Anthropological Quarterly* 55: 17–30.

Kraay, H. 1998. *Afro-Brazilian Culture and Politics: Bahia, 1790s to 1990s*. New York: M.E. Sharpe.

Landes, R. 1947. *The City of Women*. Albuquerque: University of New Mexico Press.

Martin, E.P., and J.M. Martin. 1978. *The Black Extended Family*. Chicago: Chicago University Press.

McAdoo, H.P. 1981. *Black Families*. Beverly Hills: Sage.

Medeiros, M.A. 2018. *Marriage, Divorce, and Distress in Northeast Brazil: Black Women's Perspectives on Love, Respect, and Kinship*. New Brunswick: Rutgers University Press.

Métraux, A. 1951. Brazil: Land of Harmony for All Races? *Courier* 4 (4): 3.

Mintz, S., and R. Price. 1976. *An Anthropological Approach to the African American Past: A Caribbean Perspective*. Philadelphia: Institute for the Study of Human Issues.

Mintz, Sidney W., and Eric R. Wolf. 1950. An Analysis of Ritual Co-Parenthood (Compadrazgo). *Southwestern Journal of Anthropology* 6 (4): 341–368.

Pierson, D. 1942. *Negros in Brazil: A Study of Race Contact at Bahia*. Chicago: The University of Chicago Press.

———. 1954. Família e Compadrio numa Comunidade Rural Paulista. *Sociologia* 16 (4): 368–389.

Schwartzman, Luisa Farah. 2007. Does Money Whiten? Intergenerational Changes in Racial Classification in Brazil. *American Sociological Review* 72 (6): 940–963.

Smith, R.T. 1956. *The Negro Family in British Guiana*. London: Routledge and Kegan Paul.

Stack, Carol B. 1974. *All Our Kin: Strategies for Survival in a Black Community*. New York: Harper and Row.

Telles, E.E. 2004. *Race in Another America: The Significance of Skin Color in Brazil*. Princeton: Princeton University Press.

Telles, E., and A. Esteve. 2019. Racial Intermarriage in the Americas. *Sociological Science* 6: 293–320.

CHAPTER 7

Keeping Up Appearances

Beauty and physical appearance are matters of daily concern for Brazilians from a range of backgrounds, many of whom spend considerable amounts of money and time caring for their hair, bodies, skin, and nails.[1] In the early morning rush hour traffic, women emerge from crowded buses looking feminine and fresh, hair well-groomed, smelling of perfume, and wearing clothes that reveal the bikini lines of a weekend spent sun-tanning at the beach. *Academias* (gyms) exist on almost every corner of the city, offering a wide variety of exercise and nutrition regimes for toning[2] the bodies of women and men. The city promotes healthy lifestyles by closing off part of the Orla (the main Atlantic avenue) in the mornings so that residents can walk, jog, or take a sea bath. At the popular Porto da Barra,

[1] In 2007 alone, Brazilians spent US $22 billion on hygiene products, making the country the third largest consumer of cosmetic products in the world (Machado-Borges 2009).

[2] Earlier versions of the arguments prresented in this chapter were developed in previous publications by the author. Please see: "A Beleza abre Portas: Beauty and the Racialized Body among Black Middle-Class Women in Salvador, Brazil" (July 17, 2013, *Feminist Theory* 14: 203–218); "Negotiating inequality: The Black Middle Classes in Salvador, Brazil," in *People, Money and Power in the Economic Crisis: Perspectives from the Global South*, 106–128, edited by Keith Hart and John Sharp (2014, Berghahn Press); and "Body Beautiful: Comparative Meanings of Beauty in South Africa, Brazil and Jamaica," in *Beauty and the Norm: Debating Standardization of Bodily Appearance*, edited by Sarah Böllinger, Claudia Liebelt, and Ulf Vierk (2018, Palgrave Macmillan).

D. J. Gordon, *Blackness and Social Mobility in Brazil*,
https://doi.org/10.1007/978-3-030-90765-5_7

young boys and girls practice Capoeira or swim out towards the colourful fishing boats anchored in the harbour.

In this chapter, however, I argue that practices centred on styling and managing the body have particular connotations for black middle-class women, though men are also affected. In my interviews with research participants, concerns having to do with race, beauty, and physical appearance emerged as central themes, especially in the lives of women living and working in upscale areas of the city. In the social stratum in which I moved, practices and discourses surrounding beauty and body presentation were particularly significant and underscored deeper, politicized issues about race, visibility, social mobility, and citizenship in contemporary Brazil. These struggles are racialized and gendered in ways that position the black female body as particularly problematic. Being well-dressed, elegant, and demonstrating *boa educação* (a good upbringing, being well-educated and culturally aware) can offset the effects of skin colour to some extent. For example, Helena Teixeira explained to me that she never stepped outside her door without her hair properly groomed (i.e. straightened), her skin looking good, and her clothes looking fashionable. In her 60s, she had beautiful skin and one had to look very closely to spot a wrinkle or line. Keeping up appearances, as the title of the chapter suggests, helped research participants to negotiate unequal hierarchies and avoid racism as much as possible in daily life.

Empowering symbols of black female beauty and vanity are evident in popular religious festivals celebrated throughout the year. There are instances (today and in the past) of powerful, black female religious leaders and personalities who exercise much social influence and stand out as icons of beauty in Salvador. However, black female bodies are increasingly commoditized in the city's growing tourism, entertainment, and sex industries which capitalize on Salvador's image as the most "African" city of Brazil. Images of black women appear in advertisements, pamphlets, and tourist brochures all over the city—for example, of traditional Baianas (women who sell *acarajé,* a typical street food), as spiritual leaders of Candomblé, and as dancers and singers. These images were sometimes problematic for research participants who felt that alternative images of successful and educated black Brazilians contributing to national development were few and far between. Peggy Lovell (1999, 150) has argued that an "elitist profile" of racial discrimination in Brazil may affect black women who rise to the top of the occupational hierarchy. This suggests that the sites and types of discrimination, as well as the responses to them, may differ between black lower- and middle-class women. Thus, the question arises: does the use of techniques to modify the body by black middle-class women work as a way to widen (or diminish) the gap between women from different classes/skin colours? Coming from a more powerful socio-economic position, do black middle-class women play a potentially significant role in establishing and perpetuating, or even challenging, Brazilian social hierarchies? In the following paragraphs, I explore how black middle-class women embody and experience beauty and how, through practices and discourses centred on physical appearance, they both reinforce and challenge broader social and racial inequalities.

Consuming Brazilian Modernity

We were gathered in the foyer of the Teatro Castro Alves, the city's premier theatre located in Campo Grande, to watch the first performance of Anabela Teixeira's partner, Stefano, in a ballet company. The theatregoers chatted while drinking cups of coffee or nibbling on delicious *coxinhas*—a fried, savoury delicacy usually served at parties. The women were dressed in their finest evening wear, with slim bodies and salon-straightened hair. When members of the Teixeira family walked into the foyer, they represented some of the few darker-skinned faces in the crowd, but they did not look out of place. As noted previously, they were described by their friends as a particularly stylish and sophisticated family. On this afternoon, they were dressed very elegantly, make-up carefully applied, a hint of perfume, smart handbags, high-heeled shoes, and expensive sunglasses perched on top of their well-coiffed hairstyles. In a country obsessed with beautiful, svelte, sensual bodies, they blended easily into the aesthetics of the crowd, exuding confidence, style, and glamour.

Eventually, Helena Teixeira introduced me to a tall gentleman standing beside her, "This is Alberto, do you know he was the plastic surgeon who totally reformed my breasts and stomach after the divorce? His hands are like magic, he is a great artist, I am very happy with the results." Alberto smiled and accepted the praise gracefully, lifting his cup of coffee to hers. The conversation was so matter of fact that subsequently, I asked different people about plastic surgery procedures and was surprised that it was widespread, easier to access than ever before, and no longer a practice of the elite (Edmonds 2010; Kulick and Machado-Borges 2005). In Brazil, cosmetic surgeries are accessed through private clinics or at a subsidized cost through public hospitals. A leading Brazilian plastic surgeon was quoted as saying, "plastic surgery is not only for the rich. The poor have a right to be beautiful" (Edmonds 2007a, 364). Indeed, this is important in a country that is known to celebrate beauty. In the working-class district of Liberdade, Bia's mother, Norma Sanchez, exclaimed that she had done plastic surgery on her breasts after her divorce when her children were still young, emphasizing that she felt "it was time to do something for me." In Miranda Santos' family, a niece had undergone breast reduction surgery after being on the waiting list of a public hospital for years. Her parents agreed to the surgery, explaining that she would "*fazer sucesso*" (make a success in life) and maybe marry a *fazendeiro* (rich landowner)—thus linking plastic surgery to hopes of her future mobility.[3]

In Brazil, plastic surgery is promoted by the media, where it is a matter of pride that Brazil has more plastic surgeons per capita than anywhere in the world (Kulick and Machado-Borges 2005, 127). One evening, Helena Teixeira described the operation and recovery period afterwards. She said that even as it took some weeks to recover from the surgery, she felt it was worth it after having three children. She complained that her breasts had fallen to her stomach, so she did not think she looked good in her clothes anymore. This was important to her, as she worked part time in her brother's fashion model business. In order to get back her feminine shape, she decided to invest in the surgery and was very pleased with the results and the clothes she could wear again. The openness with which people spoke about plastic surgery may indicate that to have such an operation is to show that one is able to afford it: plastic surgery procedures in private clinics, such as the one that Helena went to, can cost anywhere between US $7000 and $23,000 (Machado-Borges 2008). Thus, it can function as a symbol of status, expressing strong notions of citizenship through the ability to consume and participate in an ideal of Brazilian modernity.

In the private sector, aggressive price-cutting and flexible credit plans have made plastic surgery, or *plástica* as it is commonly called, "integral to the roster of middle class aspirations" (Edmonds 2007a, 364). There are several glossy lifestyle and women's magazines devoted to cosmetic surgery, such as *Plástica e Beleza* (Plastic Surgery and Beauty), published since 1998. An internet website, United Magazines, published data on the typical profile of people who buy this magazine. The statistics show that the majority are women; a high percentage can be classified as middle class (48.6 per cent, compared to 29.7 per cent from the upper classes and 21.7

[3] Please see previous publications by the author: "A Beleza abre Portas: Beauty and the Racialized Body among Black Middle-Class Women in Salvador, Brazil," (July 17, 2013, *Feminist Theory* 14: 203–218); and "Body Beautiful: Comparative Meanings of Beauty in South Africa, Brazil and Jamaica," in *Beauty and the Norm: Debating Standardization of Bodily Appearance*, edited by Sarah Böllinger, Claudia Liebelt, and Ulf Vierk (2018, Palgrave Macmillan).

per cent from the lower classes); most are between 35 and 45 years of age; and 79 per cent of them go shopping habitually.[4] In advertisements for different plastic surgery procedures, *plástica* is clearly promoted as an investment, as one magazine editor commented: "*Plástica é um investimento que vale a pena*" ("Plastic surgery is an investment that is worth it").[5]

The idea of plastic surgery as an investment suggests that the body can be seen as capital which one can develop, transform, and/or exchange on the market, revealing possibilities for deploying this physical capital in order to gain entrance into certain arenas of Brazilian society, represented by modernity, sophistication, glamour, and wealth. This may be particularly significant for black women because as they climb the social ladder, one is expected to transform one's physical appearance into socially acceptable, upper middle-class ideals that emphasize the submersion of "African" characteristics—hence a well-known surgical procedure is called "correction of the Negroid nose" (Edmonds 2010). Blackness and poverty are gendered in particular ways, underscored by the predominance of darker-skinned women in low-paid and low-status jobs such as domestic service (Lima 1995; Lovell 1999). This association hyper-emphasizes the desire of some black women in higher social positions to distinguish themselves from blackness symbolically associated with poverty. In Helena's case, an ability to have plastic surgery signalled to others her economic power, thus ensuring her "visibility" in the public sphere. In a study on black youth culture in Salvador, Sansone (2003, 92) suggested that "consumption has been turned into an instrument to gain civil rights (*cidadania*) and consuming (conspicuously) makes one feel like a *cidadão* (citizen)." Strong notions of national belonging and citizenship are expressed through an ability to consume and participate in an ideal of Brazilian modernity that has particular resonance for upwardly mobile black Brazilians given their previously marginalized status.

Since the 1950s and especially during the military regime (1964–1985), the Brazilian state, as part of its economic modernization plan, actively promoted consumption in order to increase the growth of the national industry. During this period, Brazil developed a sophisticated and far-reaching media network that portrayed an image of Brazil as marching towards progress, modernity, and upwardly mobile lifestyles (Dalsgaard 2004; Mello and Novais 1998). In the consumer society, buying is valued as an action that locates the consumer in the society with other citizens. For many poor Brazilians, an inability to participate in the consumer society could lead to feelings of worthlessness (Dalsgaard 2004, 145), because lack of purchasing

[4] This survey was quoted in Machado-Borges (2008, 147–148).
[5] Quoted in Machado-Borges (2008, 152).

power excludes one from recognition in the consumer society. This was powerfully revealed in the recent 2013 *rolezinhos* (little strolls), which gained visibility in cities such as São Paulo and Rio de Janeiro. *Rolezinhos* consisted of large, organized gatherings of impoverished youth in shopping malls and other semi-public spaces, where their presence in large groups was not common. As Vargas (2016, 553) has argued, these groups asserted their presence in spaces of privileged consumption associated with whiteness, declaring their right to participate in that world just like any other Brazilian citizen. These actions drew widespread condemnation from a range of groups—the police, elected officials, security advisors, the news media, and others, which raise the question: to what extent, if at all, are such spaces open to all citizens, and especially to black Brazilians? This question is symptomatic of a larger, structural, and historically rooted problem: that of the place of black people and black bodies in Brazil's national project of modernization and development (Vargas 2016).

The ideal of the tall, blonde, blue-eyed woman directly descended from Europeans appeared in the media in the form of Xuxa, a popular children's television presenter. Xuxa's programme was watched by millions of Brazilian children everyday throughout the 1980s and 1990s and was still popular when I was conducting fieldwork in Salvador. She became a national icon and one of the highest paid celebrities in the country. Many young women around the country lined up to try out for their role as one of Xuxa's assistants on the show, called *paquitas*. However, *paquitas* usually looked white with blonde hair, criteria that immediately disqualified many eager job applicants. Xuxa was rivalled by the equally popular show presented by Angélica, who also had a similar phenotypic profile: she was a young, light-skinned blonde woman with green eyes and was descended from Polish, Lithuanian, and Russian immigrants. It is only recently that African descendants have become more visible in the media, such as the well-known presenter of the programme Fantástico on the TV Globo network, Glória Maria, who has openly spoken in interviews about racism on the job and in daily life. The association of blackness with poverty, low-prestige work, negatively valorized aesthetics, and everything considered the opposite of modernity and glamour no doubt contributed to their invisibility.

Theorizing Beauty

The focus on the body, physical appearance, and the rise in plastic surgery are, of course, not specific to Brazil (Davis 2013). The emphasis on physical appearance is undoubtedly linked to a number of factors such as the body's increased visibility and objectification in late capitalist consumer society (Featherstone 1990; Featherstone and Turner 1995); the growth of the health and exercise industry (Kulick and Meneley 2005); and to the effects of the sexual revolution and the feminist movement (Bordo 1993). Brumberg (1997), for example, writes about an increased emphasis on "body projects" among American girls. Like many adults in American society, she argues that girls today are more concerned with the shape and appearance of their bodies as a primary expression of their individual identity. Brumberg argues that in contemporary America, the body is regarded as something to be managed and maintained, "usually through expenditures on clothes and personal grooming items, with special attention to exterior surfaces—skin, hair, and contours" (1997, xxi). Similarly, the producers of the popular television show, "What Not to Wear," suggest that transforming your wardrobe can also transform one's self-esteem. Commercial interests play directly to these anxieties about physical appearance, employing marketing strategies that have resulted in enormous revenues for manufacturers of skin, hair, and beauty products. However, Brazilian cultural constructions of beauty and physical appearance take on special significance because of the ways in which these ideas have been historically linked to specific constructions of race.

Feminist theories have long contributed to approaches to beauty within the social sciences. However, beauty's meaning in women's lives continues to be a problem. On the one hand, feminists have argued that beauty ideals and practices reflect patriarchal domination (Bordo 1993; Wolf 1991). These analyses stress slightly different arguments but seem to coalesce on the point that beauty practices act as a means of social control over the female body. On the other hand, beauty is also seen as a potentially pleasurable instrument of female agency (Cahill 2003; Gimlin 2002). These writers emphasize women's subjective experiences of beauty and how they might derive personal satisfaction from their body work. This view of beauty as a tool of female agency has proved illuminating for interpreting research participants' experiences. Beauty practices and discourses represent an important means through which respondents could have a constructive impact on their everyday experiences, defining a space for individual action and for circumventing traditional inequalities. Focusing on the "pragmatics of beauty"—how is beauty defined, defended, subordinated, marketed, or manipulated—(Colebrook 2006, 132)—is a particularly useful approach for examining beauty in Brazil. The racialized production of social difference in the region has historically hinged on specific notions about physical appearance. I emphasize that beauty needs to be analysed as emerging out of particular contexts, thereby challenging analyses based on an implicit Euro-American reading of the subject.[6] This chapter builds on the move in feminist theory to address beauty as an empirical question, rooted in the practical experiences emerging out of everyday life (Coleman and Moreno Figueroa 2010).

The Aesthetics of Race

Although an uncomfortable topic, the "aesthetics of race" has long been central to ideas about social difference in Brazil, as was recognized in the 1980s by the social scientist Oracy Nogueira (1985), when he argued that Brazil was characterized by *preconceito de marca* (racial prejudice based on physical appearance) and *preconceito de origem* (racial prejudice based on descent). Bahian descriptive terms reflect fine distinctions in appearance. For example, *galícia* is used to refer to someone with very fair skin, blue eyes, and blonde hair; *cabo verde* refers to a person who is lighter-skinned than someone who could be called *preto* (someone with very dark skin), with features considered to be *fino* (fine), such as thin lips, straight hair, and a narrow nose. *Sarará* refers to someone with light skin, freckles, and kinky or very curly hair (Hutchinson 1952, 29). In this folk taxonomy of colour, terms referring to darker skin colour have traditionally been negatively valued. So, for example, a person described as *sarará* is someone who has not escaped the imprint of "African" features, evidenced by the texture of the hair and/or the shape of the lips and mouth.

The racialization and stigmatization of certain physical features highlight special issues for black women, impacting on their experiences in love, marriage, and the job market, among other spheres of life (Burdick 1998). The president of a small Afro-Brazilian cultural group located in Pelourinho, Okambi, spoke about this issue in relation to her personal life:

> Racism is the biggest enemy of the black woman. It is bad for the man, but for the woman, it is worse. It is a war to be black and a woman. The society is telling you that you are ugly. We do all kinds of things to our hair, because we think our hair is ugly. When I was young, I went through every hairstyle. … I straightened it, wore braids and weaves … so I could have long hair … that is the feminine ideal. But my hair started to fall out! I went to the doctor crying, but she said, "Why are you crying? Why don't you shave your head and wear big earrings? You are beautiful!" So, I did that, and everyone—even people at

[6] Please see: Gordon, D. 2013. "A Beleza abre Portas: Beauty and the Racialized Body among Black Middle-Class Women in Salvador, Brazil." *Feminist Theory* 14: 203–218.

> work—told me I looked beautiful. I know of another case, a girl who grew up in a white neighbourhood and had no boyfriend. She had white friends, but she did not have a boyfriend. She did a treatment for the nerves (nervous breakdown). She didn't accept herself … she thought she was ugly, and that is why she had no boyfriend, and she started to think she was stupid, too. She went to therapy. Now she has a fiancé. You have to assume your identity—regardless of the hairstyle you have. (Interview: June 26, 2017)

The Brazilian Black Movement has worked hard to invest previously stigmatized terms with positive meaning and to celebrate the beauty of black bodies and physical features, as in the tune composed by Macau for the popular soul singer Sandra de Sá in 1994, called Olhos Coloridos (Coloured Eyes). In this song, the singer expressed pride in her physical features, such as her darker skin colour and curly hair. More recently, Afro-Brazilians have begun to organize themselves in marches to affirm the beauty of curly or kinky black hair, in the Marcha do Empoderamento Crespo (March for the Empowerment of Curly Hair), held annually since 2015. Pre-march activities typically involve visits to schools, universities, and various neighbourhoods where activists and musicians spread positive messages about the beauty of black hair, countering dominant aesthetics that have long devalued black bodies and physical features in Brazil.

Nevertheless, there is a cultural logic of race that exists in Brazil that celebrates mixture and eroticizes brownness. This was not always the case in the past. Ideas of biological racism, emanating from Europe and prominent in Brazil until the early twentieth century, condemned the country to degeneration because of the high degree of racial mixture in the population. Some of these explanations focused on beauty and physical appearance—for example, Arthur de Gobineau, who spent 15 months in Brazil as France's envoy, commented that "we are dealing with a totally mulatto population, corrupt of flesh, and frightfully ugly" (Schwarcz 1993, 5). These ideologies had various strands but converged on the idea of European superiority and beauty (Jarrín 2010; Wade 1997). Thus, the sexual union of two people can be perceived as a kind of "aesthetic lottery" in which offspring may have any combination of features which may or may not combine to form a symmetrical and harmonious ideal of beauty. Relatives may therefore exhibit different permutations of racial features, skin colour, and hair texture—leading sometimes to "ugly" combinations (Edmonds 2007a, 373).

For those who believed that it was impossible to achieve a civilized state with a predominantly mixed race population, the solution presented itself in the promotion of European immigration schemes, the aim being to "whiten" the population (Skidmore 1999). This discourse of "whitening" or "improving" the race rears its head at different times, even recently in the lead up to the elections in Brazil, when Hamilton Maurão—a candidate for vice president and member of the conservative party, the Brazilian Labour Renewal Party (PRTB)—attributed the beauty of his grandson to *branqueamento de raça* (whitening of the race).[7]

Given that the majority of Brazil's population remained far from white, a more realistic solution for national elites was to celebrate and promote racial mixture, portraying to the world a modern country that had overcome racial tensions. Writers such as Freyre (1984) underscored the *erotics* and the *aesthetics* of impurity, of sexual intimacy across racial lines, leading to the creation of a unique Brazilian meta-race. Freyre embraced the beauty of *morenidade* and brownness, especially in relation to the female body, resulting in a corporeal ideal of beauty celebrated in Brazil. For example, the meteoric rise of Brazilian celebrity Andressa Soares in recent years, whose stage name is Mulher Melancia (Watermelon Woman), epitomizes this national obsession with the mixed race woman. A dancer, singer, and model known for her unique style of dance, Mulher Melancia possessed the desirable traits of the *morena*: tanned skin, long, flowing hair, large rounded bottom; and the ability to dance in an erotic manner.

The Brazilian obsession with racial mixture presents an alternative scenario for feminist theory. While studies on the relationship between "race" and beauty exist (Banet-Weiser 1999; Banks 2000) feminist writers have tended to emphasize the dominance of Euro-American beauty standards that reinforce the desirability of fair skin, small noses and lips, and long hair. While this discourse is present, the national ideology of racial mixture and *morenidade* in Brazil challenges these assumptions.

Nevertheless, the national ideology of race mixture has not replaced hegemonic ideas of beauty that emphasize whiteness. Whiteness is displayed through varied modes in popular culture such as the mass media, magazines, and newspapers and is hugely influential at events such as the São Paulo Fashion Week, the nation's most pivotal event in fashion and beauty after carnival. This event has created controversy in the past over the limited number of black models in their parades, leading to federal prosecutors declaring that 10 per cent of models in this event must be of

[7] Marques Travae, "White is Right? General Maurão, an Indian and Candidate for Vice President of Brazil, Comments on How Handsome His Grandson Is and that He is An Example of the 'Whitening of the Race.'" Black Brazil Today, October 9, 2018. Retrieved on August 4, 2021 at: https://blackbraziltoday.com/white-is-right-general-mourao/

African or indigenous descent. In 2009, the prosecutor Deborah Kelly Afonso was quoted in the *Folha* newspaper as stating:

> The percentage of black models at the event … is far less than that of white models. The objective of [the public department] is to come to a consensus on social inclusion, to establish the minimum number of black models to participate in the shows.[8]

The hegemonic ideal of whiteness was also revealed in respondent's stories. For example, Nubia related an incident that occurred in primary school in which the ideal of whiteness was reinforced. Nubia was regarded by many friends as being a very attractive girl as she had smooth brown skin, long, curly hair, and unusual green eyes. One day, she entered a beauty contest at her school which they held every year, called Rainha do Milho (Queen of the Corn). She gathered the most votes from classmates but Nubia heard the teachers whispering that they could not allow a black girl to win the beauty title. All the previous Rainhas had at least looked white. So, while *morenidade* and race mixture are national ideals and have been a dominant theme in the re-imagination of the nation in the twentieth century, in particular situations the supremacy of whiteness is affirmed.

The old colour hierarchies illustrated above reappear in Salvador's contemporary, competitive work market, in the hyper-emphasis on *boa aparência* (good appearance).[9] Traditionally, *boa aparência* meant having lighter skin combined with certain signs of a higher social class position. This concept has particular meanings for female employees in the modern service and financial sectors (Edmonds 2010; Telles 2004). Employees are expected to reflect an image of modernity, cosmopolitanism, and glamour associated with city centre lifestyles. Varying in shade from dark brown to light, shop assistants, cashiers, and salespersons are mostly slim, smooth-skinned, and young, with straight or long hair.

In tension with discourses on whitening/modernity and racial mixture is the discourse of blackening (see Chap. 3). Since the 1970s, black feminists have brought attention to issues of black women's sexual and

[8] Folha Online, "Promotora Quer Cota para Negros em Desfiles de Moda," April 12, 2009. Retrieved on August 3, 2021 at: https://www1.folha.uol.com.br/cotidiano/2009/04/549446-promotora-quer-cota-para-negros-em-desfiles-de-moda.shtml

[9] The Spanish term, *buena presencia*, the corollary of the Brazilian Portuguese term, *boa aparência*, is mentioned in Sutton (2010) who writes on Argentina, and Cassanova (2011) who writes on Ecuador.

reproductive health as well as sexual, racial, and domestic violence issues (Caldwell 2007; Santos 2008). Black women's organizations are involved in efforts to build self-esteem among black women through workshops, self-help groups, and community outreach. For example, in the 25 years since its creation in 1993, thousands of girls and women have passed through the Didá Educational and Cultural Association of Salvador, carving out a space for black women in the percussion community. In an interview with Víviam Caroline de Jesus Queirós (Cultural Director of Didá), she highlighted the work she does in building the self-esteem of black girls and women, especially in the area of beauty and physical appearance:

> I get together every Wednesday with the girls to approach the theme of black beauty and self-esteem by dialoguing with them. I believe that when we dance and play the drums and create our hairstyles and do our makeup, we create an elaborate aesthetic, we are doing it in a very natural way while developing the girls' self-esteem … we are developing this power of affirmation and belonging. (Interview: June 29, 2017)

Similarly, in the Afro-bloco Malê Debalê, the Coordinator of Malezinho, the children's programme of the bloco, commented on her work in the Itapuã community, especially with regards to strengthening concepts of black beauty among youth:

> We work with children in the community by strengthening not only the ideal of black beauty but also reaffirming the fight against prejudice and racism. We know that these feelings cause deep wounds that almost don't seem to heal; so, it isn't fair that our children continue experiencing and growing with this. So, we do this social program, taking children off the streets and out of situations of possible marginality and working with them using dance and music, therefore working with their self-esteem. This makes them accept themselves and recognise themselves as black children. They need to look in the mirror and like what they are seeing. Children need to look at themselves and think, "Wow, I am beautiful. My hair is beautiful. I am marvellous." Years ago, children didn't have this: they didn't accept themselves, by saying they were not black, claiming to be *moreno* and saying that their hair was not hard. (Interview: June 27, 2017)

Workshops on hair braiding underscore efforts to challenge dominant prejudices regarding the hair texture of black women (Caldwell 2007, 159). As Víviam Caroline de Jesus Queirós commented, "The hairstyle work is important. The hair is central to the question of black beauty." Similarly, the Coordinator of Malezinho commented that:

> We did a whole program with these children so that they would understand that the term "hard" as a descriptor of black hair reflects prejudice from certain human beings towards black bodies. Black women's hair isn't "hard," it's kinky and curly. It's the hair of black people. So, we have to look at the issue of ancestrality ... the hair and body type of our ancestors ... that these were people that were enslaved, but they were not born slaves. We need to accept ourselves as black. Because if we do not recognise and accept ourselves as black, then people in the society will not accept us. So, we need to affirm ourselves, to say that, yes, I am black, my hair is kinky, and I love my hair. (Interview: June 27, 2017)

These efforts have had varying effects. In some cases, the organizers and leaders of black cultural groups feel that they are achieving their objectives and that children are much more willing to accept themselves as black, especially by being involved in their culture through music and dance where they develop a sense of pride. However, they also point to cases where the parents of the children were not happy about their changed appearance once they entered the group. For example, Víviam recalled that one parent refused to have her daughter appear in a media interview with her hair in an afro. On the other hand, the popularity of blonde hair extensions among women who are the target of these efforts underscores the need to evaluate more closely conditions of power in the lives of lower- versus middle-class women. Figure 7.1 is a picture of the author, who was invited to an Afro-Brazilian hair-braiding workshop in Liberdade.

The increased promotion of beauty practices that express pride in blackness has challenged hegemonic models present in Brazilian society that have tended to highlight lighter skin colour and hair. An expanding black beauty industry (occurring in Brazil and elsewhere) seems to have arisen to address this need (Santos 2000; Gomes 2006). The contemporary concept of *beleza black* (black beauty)—which derives from African American civil rights struggles during the 1960s—is associated with these Black Movement discourses. It also reflects links between the Black Movement in Brazil and a more broad-based pan-African movement (Sansone 2003). Cultural organizations in the city emphasize blackness as a source of pride and beauty as their members march through the streets in African-style fashion featuring bold prints and colours, their hair in braids and decorated with bright beads.

However, the situation is highly complex: there are competing hierarchies of beauty in which understandings of what is "white," "brown," or "black" resists a definitive reading. What is hegemonic in one space (e.g. black aesthetics is dominant at Ilê Aiyê's rehearsals in Liberdade) is not

Fig. 7.1 The author in an Afro-Brazilian hair-braiding workshop, Salvador. (Source: Doreen Gordon)

hegemonic in another space (e.g. in the luxurious shopping malls where whiteness is dominant). A person may also embody "brownness" in one social context but in another, their physical appearance may be classified as white (e.g. a light-skinned person in a space where the population is overwhelmingly dark-skinned). It is this very flexibility that allows for racial boundary crossing which has been implicated in processes of social mobility and whitening (Schwartzman 2007). Indeed, the possibility of "passing" from one racial group to another creates the conditions whereby subtle cosmetic changes could nudge a person closer to *boa aparência*—a word that is often synonymous with looking more European (Edmonds 2007b). Therefore, the idealization of inter-racial mixing and hybridity that Brazil is famous for coexists with historically entrenched aesthetic[10] hierarchies.

[10] Earlier versions of these arguments were developed in a previous publication by the author. However, substantial new evidence has been included here. Please see: "A Beleza abre Portas: Beauty and the Racialized Body among Black Middle-Class Women in Salvador, Brazil." (July 17, 2013, *Feminist Theory* 14: 203–218).

Let's Talk About Her Hair

Hair texture is regarded as one of the most important, visible signs of the social-aesthetic power of a woman and is related to culturally meaningful symbols of femininity (Machado-Borges 2008; Malysse 2002). Long, wavy hair or salon-straightened hair—associated with a "European" appearance—is generally preferred. Meanwhile, tightly curled, kinky hair—associated with an African heritage and defined as socially ugly and unfeminine—is often referred to as *cabelo ruim* (bad hair), *cabelo duro* (hard hair), and *cabelo feio* (ugly hair). Indeed, it was not long ago when, in July 1996, an album was released that included a song, "*Veja os cabelos dela*" (Look at her hair) by Tiririca, a Brazilian singer, comedian, and politician. While popularly regarded as playful and humorous, the song included several negative stereotypes about black women, highlighting the gendered dimensions of racism. The song refers to a woman who initially captures the singer's attention, but whose hair he finds undesirable, likening it to a scouring pad used to clean pots and pans. His description of her also makes reference to her smell and skin colour, such as:

> *Essa nega fede, fede de lascar.* (This black woman stinks, she stinks horribly.)
>
> *Bicha federenta, fede mais que gamba.* (Stinking beast, she smells worse than a skunk.)

The song associates black bodies with ugliness and stench—and in particular, stigmatizes the texture of black women's hair. The lyrics created an outcry among black activists and resulted in both Tiririca and his recording label, Sony Music, being sued for racism. However, the song was not seen as offensive by most Brazilians, but rather as a *brincadeira* (joke). Indeed, Brazilian discourses of race have frequently used humour in order to transmit racist stereotypes (Goldstein 1999; Sheriff 2001). Popular acceptance of racist humour in Brazil indicates that joking often provides a culturally sanctioned means of articulating ideas that represent dominant notions of white superiority and black inferiority.

Such negative valorizations of black bodies have contributed to an ongoing stigma and sense of low self-esteem among black youth, though people seldom accept these notions without resisting or defining their meaning (Santana Pinho 2006). For example, every year Ilê Aiyê offers a course, Beleza Negra (Black Beauty), for the girls and women who want to be Queen of their carnival parade. In an interview with the coordinator of this programme, she explained that as she prepares them for the Beleza Negra competition, she must work on raising their self-esteem in being black:

> Black women have been very much discriminated against, especially black hair. The black woman has a broad nose, and thick lips, for example … a body that is rounder. Today we have slimmer girls as well. But this does not concern us. What concerns us is that they have an elevated self-esteem. Once, there was a girl who arrived here who was so pretty, but she did not think so. The girls who do the course often resist the idea the whole way. They think other girls are beautiful. There was another case where the girl did not show up for the competition and she called me to say sorry. And I had to say, "Wait a little, let us talk." Sometimes, the psychologist comes to talk with them. We must tell them that if you do not wake up and think you are beautiful, then those people around you will break your spirit with their opinion. It is a lot of work. All the time, we must push them and encourage them. (Interview: June 27, 2017)

Similarly, Víviam, Cultural Director of Didá, stated that:

> I always talk to the girls and tell them that "you are beautiful, accept yourself, your hair is marvellous, love it." Because there are many children who come here with this feeling of inferiority. This feeling that they are not beautiful. This prejudice that their hair isn't nice. There are some people who may make fun of them. For some of these children, I had to basically build this identity. Because they would cry and writhe and scream, "I am not black, I am *morena*. My hair is straight!" So, then I had to pull them aside and say, "Look here, my love, *morena* is a white woman with black hair; a woman of fair skin. You have black skin. You are beautiful. Your hair is kinky. It is hair that belongs to us black people." Then she would stop, she would look about and say, "ahh my pro (professor/teacher), I don't like this idea, you know." Then I would tell her to let me comb her hair and show her how awesome her hair is. Then I did some *biotes* (also called pompoms or bantu knots). I put on some lipstick and I told her, "Look into the mirror, are you beautiful or not?" She said "Aii pro (professor/teacher), I am feeling myself." So, I needed to show her that she was beautiful. (Interview: June 27, 2017)

Nevertheless, the idea of at least taming the black characteristics of the body took on added significance for some upwardly mobile black people. Straightening the hair (through hairstyling or the application of chemicals) was one such beauty practice that could modify appearance. The link between straight hair and a kind of raced and classed femininity (closely associated with a "European" appearance) was enacted by participants' beauty practices but not explicitly stated. One of the more popular hair techniques was the *alisamento Japônes* (Japanese straightening), where hair is blow dried until it supposedly looks like the shiny hair of Japanese women. Other practices involved the application of creams and gels to the hair so that it came out long and wavy—in keeping with feminine ideals of beauty.

The politics of race in Brazil has led to moral arguments that those with straightened hair want to cultivate a "European" appearance and are therefore suffering from low self-esteem and a mutilated black consciousness. Some black activists insist that aesthetic affirmation is necessary to overcome internalized racism. Yet participants asserted their right to style themselves in whatever way they felt like without being judged for it. Research participants might reject, or experiment with, practices that emphasize an "African" aesthetic, such as styling the hair in braids or dreadlocks. They stress that they have the right to choose how they want to look. From a post-feminist perspective, these participants might be seen to be adopting de-politicized stances towards social inequalities in their country. In *The Aftermath of Feminism*, Angela McRobbie argues that consumer culture in contemporary society offers the illusion of "freedom of choice" when in reality, new forms of power are enacted that re-establish patriarchal hegemony and the cultural dominance of whiteness. However, elements of creative play involved in participants' beauty work are neglected in this perspective. Applying blonde highlights or implanting long hair weaves are also ways of inverting dominant beauty standards. In other words, some participants resisted their positioning by others to a certain racial category or appearance (Malysse 2002). On other occasions, however, blonde hair could be "read" differently according to who is wearing the style. For example, one informant, who is dark-skinned and now living in a middle-class neighbourhood with his mother, invited some friends over to his house. When his friends began to tease him about his mother's blonde hairstyle, he was mortified. This signifies that beauty practices are also ranked in terms of social prestige.

Hair was clearly a very delicate subject for participants who lived, worked, or went to school in areas associated with whiteness. Nubia recounted that learning to care for her daughter's hair was one of the most important identity-forming experiences of her life. Nubia had always identified with being *morena*, due to her unusual green-brown eyes, wavy brown hair, and golden-coloured skin. Her daughter is darker-skinned, the product of a first marriage. When this marriage ended and she moved to a new neighbourhood, she encountered problems with her daughter's self-esteem as she was one of the few darker-skinned pupils at the prestigious school she attended. Her daughter went through painful experiences that in large part had to do with the texture of her hair. She was ridiculed, teased, and made to feel unattractive by her peers. Concerned about her daughter's confidence, Nubia tried to find ways of taking care of her hair, showing her daughter different hairstyles she could try, make-up that would complement her dark skin, fashion styles she could experiment with, all the while telling her that she was beautiful just the way she was. Today, Nubia's daughter is a successful model in São Paulo, appearing in fashion events and commercials advertising credit cards and mobile phones. It is questionable whether she would have had this kind of success in Salvador, even though it is the city most associated with a celebration of blackness.[11] Nubia's struggles with her daughter's racial identity led to Nubia's strong identification as *negra* (black), despite the protests of close friends and family, for whom she was *morena* (brown/mixed race).

The idea of un-straightened hair has gained currency among some participants, especially with the increased availability of hair-care products now on the market. One informant, Ana-Paula, preferred products from the United States and would ask her American friend to bring several packs of organic afro-hair extensions and special gels which she could use to create various hairstyles. These included hairstyles such as the afro, braids, and dreadlocks, which signify the embrace of an alternative ideological code of value. Some professional women—like educators, doctors, and lawyers—wore their hair this way. However, it should be noted that women in a more powerful, socio-economic position perhaps have more options to challenge the racial status quo than women in low-income communities whose options for social mobility are more constrained.

Raça Brasil and the Black Consumer

Raça Brasil, a new magazine on the market targeting a largely black middle-class clientele, has several pages dedicated to the advertisement of beauty products that claim to nurture black skin and hair, with pictures of African American and local celebrities who use them. Indeed, there has been an increase in ethnic salons in Brazil in recent years (Santos 2000;

[11] The systematic racial discrimination that African descendants faced in São Paulo's competitive job market appears to have fostered a more radical posture to "race" and political activism in that city. In Salvador, selective avenues of upward mobility allowed for the presence of African descendants in higher social positions, leading to a more conservative attitude to race (Butler 1998). This could have made São Paulo a more progressive environment for Nubia's daughter.

Gomes 2006). The magazine *Raça Brasil* was created with a vision in mind to counteract the negative stereotyping and racial prejudice towards the phenotype and hair of people of African descent. When the magazine was first produced in 1996, the editor promised to *"Dar a você, leitor, o orgulho de ser negro. Todo cidadão precisa desse dose diaria de auto-estima: ver se bonito, o quatro cores, fazendo successo, dançando, consumindo. Vivendo a vida feliz"* (We aim to give to you, the reader, pride in being black. Every citizen needs this daily dose of self-esteem: to see themselves as beautiful, the four colours, making a success, dancing, consuming. Living a happy life.)[12] The producers of the magazine felt that there was a good market in Brazil for a black magazine such as *Ebony* in the United States. Wanting to appeal to as wide a cross-section as possible, the editors decided not to take a black militant position but rather to "*vender a imagem de um negro diferente, que não era vendido ate então*" (sell an image of the black person that had never been sold before until now).[13] A large portion of the magazine is dedicated to black people who appear successful in life and includes make-up tips, fashion, hair, and cosmetic products. This can be linked to an earlier point made about consumption and status in Brazil.

In the post-war period, especially between the years 1950 and 1979 when Brazil experienced accelerated growth and heightened consumption, the vision of Brazil marching towards progress and modernity that was emphasized by the media and the military dictatorship both implicitly and explicitly associated this modernity with whiteness (Mello and Novais 1998). New patterns of consumption emerged associated with rapidly changing lifestyles especially in large urban centres. These included the development and proliferation of fast food restaurants; large supermarket chains and clothing department stores; new imported and locally produced household, food, and electrical products; and cosmetics (Mello and Novais 1998; O'Dougherty 2002). However, the inequality in the Brazilian population was also great which meant that while consumption was symbolically important, it was also out of the reach of many. However, as a small, black middle class began to expand, the need to portray Afro-Brazilians as consumers, just like any other citizen, became apparent. Advertisements and commercials increasingly portrayed African descendants using mobile phones, computer technology, and credit cards (Fry 2002).

[12] Quoted in Fry (2002, 314).
[13] Ibid.

Work, Sexuality, and *Boa Aparência*

An excursion to the various shopping malls around Salvador offers a varied experience—even the smells and fragrances in the malls are different, perhaps due to the different foods available for sale and the use of air conditioning or air fresheners. Shopping Piedade and Shopping Liberadade, for example, are known as shopping centres for *o povo*, or the masses, whereas Shopping Barra, Shopping Iguatemi, and, later, Shopping Salvador are upmarket malls (with upmarket prices) in some of Salvador's premier neighbourhoods. Shopping Barra offers a secure, comfortable, and air-conditioned experience in which a mostly middle-class clientele come to socialize with friends and family, browsing the shops, eating a meal, or going to the cinema. Whenever I entered Shopping Barra, I saw a particular image near the entrance—a large poster of a dark-skinned, slim young lady dressed in high heels, wearing a blouse and skirt, with long, flowing hair, a cellular phone in her hand, and a pair of dark glasses resting on top of her head. The image represented the modern lifestyle of the busy career woman, with the freedom to spend money and wear what they wanted. This image of success, social mobility, and glamour had a black face, which certainly challenged traditional images of African descendants in other places around the city where they are represented mostly as Candomblé practitioners or serving up delicious plates of Afro-Bahian food. Here was an image of a black person positioned within an urban, cosmopolitan setting with the kind of lifestyle associated with whiteness and wealth.

Significant changes and restructuring of the labour market in Salvador may have created the conditions for a hyper-emphasis on appearance. The emergence of a feminized service industry (also called "pink collar" workers) is not specific to Brazil: similar changes have occurred in the labour market in other countries as far apart as Ireland, the Dominican Republic, Jamaica, Mauritius, and the United States (Freeman 1993). Santos (1998) states that between the years 1987 and 1999 in the Metropolitan Region of Salvador, there was an expansion of the service and commercial sectors at the same time that there was a contraction in jobs generated in the formerly more productive sectors such as civil construction. In addition, the commercial sector underwent significant restructuring, with an increase in small commerce and a reduction in jobs protected by social security. While whites have been able to maintain their relative presence in the more structured segments of commerce, blacks have been more exposed to the vulnerability and precariousness of work (Santos 1998, 143).

Women have steadily increased their participation in the labour force, especially in personal services, health and educational services, finance, and insurance. However, black women tended to be overrepresented in low-paid occupations such as domestic work and nurse aides—nevertheless, due to the enforcement of minimum wage laws, their incomes likely improved (Telles 2004, 124). These kinds of transformations may have contributed to a special emphasis on physical appearance for women, especially in jobs that require direct contact with customers and in situations where the job market is precarious, unstructured, and highly informal. The jobs on offer are not necessarily better paid—but they are associated with the glamour and sophistication of modern, trendy, urban lifestyles, which act as a powerful magnet for young people. In addition, the emphasis on dressing professionally implies that the way one looks reflects one's work ethic, productivity, and ability to please the client, thus underscoring powerful metaphors of "corporate discipline" as well as individual expression and pleasure in the way that one dresses (Freeman 1993, 177). Being just a little closer to consumer products also enhances the idea that one is participating in a consumer society. Anita Teixeira, who had previously worked in this sector selling life insurance in a bank, commented on the lifestyle associated with this type of work:

> I liked working in the bank. It had a certain status. You have to leave your house looking beautiful and well-groomed … there is this as well, right? I liked this aspect of working there. I had an allowance to buy clothes. But I left the bank because it wasn't paying me that much and I wasn't getting anywhere. I didn't have a formal position, they did not sign my *carteira* (work booklet), I didn't get reimbursed for transport or for lunch, I didn't get holidays. I was working almost 3 years without any holiday, so now I am unemployed. (Interview: April 16, 2007)

Since the time of this interview, Anita has significantly improved her situation by enrolling in university and getting her Bachelor's and Master's in Environmental Engineering: she now lives and works in Chapada Diamantina, in the interior of Bahia. However, in reflecting on the lifestyle she used to have as an insurance salesperson, she liked its association with modernity, sophistication, and glamour. In this environment where employees interact directly with the customer, *boa aparência* and traits such as youth and sexual allure "add value" to the service interaction, while colour continues to structure access to some types of jobs (Edmonds 2007a, 369; Hasenbalg and Silva 1999).

In highly competitive situations where job markets are increasingly unstructured and informal and where applicants possess similar educational credentials, what might give a particular person the edge is his/her presentation of self, including physical appearance. These conditions present fertile grounds for anxieties about appearance, for both women and men. Indeed, as Santos (1998) suggests, men may increasingly be finding work in the informal sector—such as in the fitness or health industry, where investment in physical capital (the body) is important and can be converted into economic and social capital.

Broad changes in gender and sexual/affective relations have also occurred. For example, divorced and older women and men re-enter the dating/marriage market. Some do not feel that you have to accept that you are getting old. As Helena remarked, "An older woman can look pretty good with a few nips and tucks here and there, and if she takes care of her skin and body." Helena, as a divorced woman, was also dating at the same time that her younger daughter, Anita, was dating. Therefore, sex is being legitimated for new groups of women—the middle aged, the divorced, and the adolescent (Edmonds 2007a; Goldenburg 2004). Cultivating the body is therefore increasingly important. The use of body techniques—such as plastic surgery, diet pills, and exercise—promotes a sense of modern, "hygienic" bodies in tandem with the image of a rapidly developing Brazil.

These broad social transformations may create the sensation that the body is freed of its traditional constraints and create feelings of integration and belonging to a society, where upwardly mobile African descendants can be a consuming citizen just like any lighter-skinned Brazilian. In this context, beauty practices may present ways of remaining competitive in work, sexual, and affective markets. Here, the management and development of the body is central in its own right to human agency in Salvador.

Discussion

Beauty and a pre-occupation with physical appearance have long been associated with mass culture, femininity, frivolity, and imposed consumption in academic debates (Machado-Borges 2009, 208). The potentialities of beauty as a tool of social change have received less attention in the literature on race relations in comparison to more radical processes, for example, studies of social movements (Burdick 2005; Hanchard 1994). Yet the material in this chapter underscores the ways in which respondents engaged with Brazilian racial hierarchies in potentially revolutionary ways. However, the contradictory position of black middle-class women—as representing both a marginalized and an economically better-off stratum of the population—poses some complex issues for interpretation. Notions of femininity in particular are tied to traditional ideas that privilege a "European" appearance. However, recent shifts in racial thinking in Brazil have made it less problematic to identify with blackness. Alternative notions of beauty emanating from the West—such as the African American concept of black beauty—have mixed with local Black Movement discourses to produce a more positive valorization of blackness. Some respondents embraced these discourses, while others resisted being classified as belonging to any particular racial group when it came to their beauty practices. They asserted the right to choose from a variety of options, which is suggestive of the need to further examine the impact of rising consumer culture and neoliberalism on local subjectivities. Writers also analyse the contemporary dynamics of post-feminism such as McRobbie (2009), who suggests that consumer culture appeals to ideas about female success and choice while it in fact reinforces traditional power dynamics and ties women into new, neurotic tendencies such as eating disorders. In the case of research participants, consumption through the body (fashion, diets) and consumption around the body (beauty treatments, surgery), represented symbolic and material ways of positioning themselves within contemporary Brazilian society—thus becoming "visible" and therefore a citizen.

References

Banet-Weiser, S. 1999. *The Most beautiful Girl in the World: Beauty Pageants and National Identity*. Berkeley: University of California Press.

Banks, I. 2000. *Hair Matters: Beauty, Power, and Black Women's Consciousness*. New York: New York University Press.

Bordo, S. 1993. *Unbearable Weight: Feminism, Western Culture, and the Body*. Berkeley: University of California Press.

Brumberg, J.J. 1997. *The Body Project*. New York: Random House.

Burdick, J. 1998. *Blessed Anastacia: Women, Race, and Popular Christianity in Brazil*. New York: Routledge.

———. 2005. Why Is the Black Evangelical Movement Growing in Brazil? *Journal of Latin American Studies* 37: 311–332.

Cahill, A.J. 2003. Feminist Pleasure and Feminist Beautification. *Hypatia* 18 (4): 42–64.

Caldwell, K.L. 2007. *Negras in Brazil: Re-envisioning Black Women, Citizenship, and the Politics of Identity*. New Brunswick: Rutgers University Press.
Colebrook, C. 2006. "Introduction." Special Issue on Beauty, Race and Feminist Theory in Latin America and the Caribbean, edited by Monica Moreno Figueroa and Megan Rivers-Moore, *Feminist Theory* 7 (2): 131–142.
Coleman, C., and M. Moreno Figueroa. 2010. Past and Future Perfect? Beauty, Affect and Hope. *Journal for Cultural Research* 14 (4): 357–373.
Dalsgaard, A.L. 2004. *Matters of Life and Longing*. Copenhagen: Museum Tusculanum Press.
Davis, K. 2013. *Reshaping the Female Body: The Dilemma of Cosmetic Surgery*. New York: Routledge.
De Cassanova, E.M. 2011. *Making Up the Difference: Women, Beauty, and Direct Selling in Ecuador*. Austin: University of Texas.
De Santana Pinho, P. 2006. Afro-Aesthetics in Brazil. In *Beautiful/Ugly: African and Diaspora Aesthetics*, 266–289. Durham: Duke University Press.
Dos Santos, L.C.C. 1998. Desigualdades Ocupacionais entre Brancos e Negros Na Região Metropolitana de Salvador (1987–1997). Master's thesis, Universidade Federal da Bahia.
Dos Santos, J.T. 2000. O Negro No Espelho: Imagens e Discursos Nos Salãos de Beleza Étnicos. *Estudos Afro-Asiáticos* 38: 49–65.
Dos Santos, S.B. 2008. Brazilian Black Women's NGOs and Their Struggles in the Area of Sexual and Reproductive Health: Experiences, Resistances, and Politics. PhD dissertation, University of Texas.
Edmonds, A. 2007a. The Poor Have a Right to Be Beautiful: Cosmetic Surgery in Neoliberal Brazil. *Journal of the Royal Anthropological Institute* 13: 363–381.
———. 2007b. Triumphant Miscegenation: Reflections on Beauty and Race in Brazil. *Journal of Intercultural Studies* 28 (1): 83–97.
———. 2010. *Pretty Modern: Beauty, Sex, and Plastic Surgery in Brazil*. Durham: Duke University Press.
Featherstone, M. and B.S. Turner. 1995. "Body and Society: An Introduction." *Body and Society* 1 (1): 1–12.
Featherstone, M., ed. 1990. *The Body: Social Process and Cultural Theory*. London: Sage.
Folha Online. 2009. Promotora Quer Cota para Negros em Desfiles de Moda, April 12. https://www1.folha.uol.com.br/cotidiano/2009/04/549446-promotora-quer-cota-para-negros-em-desfiles-de-moda.shtml. Accessed 3 Aug 2021.
Freeman, C. 1993. Designing Women: Corporate Discipline and Barbados' Offshore Pink-Collar Sector. *Cultural Anthropology* 8 (2): 169–186.
Freyre, G. 1984. Uma Paixão Nacional. *Playboy*, 113 (December). The following link is no longer available: http://www.releituras.com/gilbertofreyre_bunda.asp. However, the article is extracted on the following website, "Conto

Basileiro: Contos, Crônicas e Poesias de Autores Brasileiros." https://conto-brasileiro.com.br/bunda-paixao-nacional-artigo-de-gilberto-freyre/. Accessed 6 Aug 2021.

Fry, Peter. 2002. Estetica e Politica: Relacoes entre Raca, Publicidade e Produção da Beleza no Brasil. In *Nu e Vestido: Dez Antropologos Revelam a Cultura do Corpo Carioca*, ed. Miriam Goldenberg. Rio de Janeiro: Editora Record.

Gimlin, D. 2002. *Body Work: Beauty and Self Image in American Culture*. Berkeley: University of California Press.

Goldstein, D. 1999. 'Interracial' Sex and Racial Democracy in Brazil: Twin Concepts? *American Anthropologist* 101 (3): 563–578.

Gomes, N.L. 2006. *Sem Perder a Raiz: Corpo e Cabelo Como Simbolos da Identidade Negra*. Belo Horizonte: Autentica.

Hanchard, M.G. 1994. *Orpheus and Power: The Movimento Negro of Rio de Janeiro and São Paulo, Brazil, 1945–1988*. Princeton: Princeton University Press.

Hasenbalg, C.A., and N. Do Valle Silva. 1999. Race and Educational Opportunity in Brazil. In *Race in Contemporary Brazil: From Indifference to Inequality*, ed. R. Reichmann. Pennsylvania: The Pennsylvania State University.

Hutchinson, H.W. 1952. Race Relations in a Rural Community of the Bahian Recôncavo. In *Race and Class in Rural Brazil: A UNESCO Study*, ed. Charles Wagley, 16–46. New York: Columbia University Press.

Jarrín, A. 2010. Cosmetic Citizenship: Beauty, Affect, and Inequality in Southeastern Brazil. PhD dissertation, Duke University.

Kulick, D., and T. Machado-Borges. 2005. Leaky. In *Fat: The Anthropology of an Obsession*, ed. Don Kulick and Anne Meneley, 121–137. New York: Jeremy P. Tarcher/Penguin.

Kulick, D., and A. Meneley, eds. 2005. *Fat: The Anthropology of an Obsession*. New York: Penguin Group.

Lima, M. 1995. Trajectoria Educacional e Realização Sócio-Econômica das Mulheres Negras. *Estudos Feministas* 2: 489–495.

Lovell, P. 1999. Women and Racial Inequality at Work in Brazil. In *Racial Politics in Contemporary Brazil, edited by Michael Hanchard*. Durham: Duke University Press.

Machado-Borges, T. 2008. O Antes e Depois: Feminilidade, Classe, e Raça na Revista Plástica e Beleza. *Luso Brazilian Review* 45 (1): 146–163.

———. 2009. Producing Beauty in Brazil: Vanity, Visibility and Social Inequality. *Vibrant* 6 (1): 208–237. http://www.vibrant.org.br/issues/v6n1/thais-machado-borges-producing-beauty-in-brazil/. Accessed 7 Aug 2021.

Malysse, S.R. 2002. Extensões do Feminino: Megahair. *Baianidade e Preconeito Capilar. Studium* 11. https://econtents.bc.unicamp.br/inpec/index.php/studium/article/view/11739. Accessed 7 Aug 2021.

McRobbie, A. 2009. *The Aftermath of Feminism: Gender, Culture and Social Change*. London: Sage.

Mello, J., and F. Novais. 1998. Capitalismo Tarde e Sociabilidade Moderna. In *Historia da Vida Privada No Brasil*, ed. L.M. Schwarcz. São Paulo: Companha das Letras.

Nogueira, O. 1985. *Tanto Preto Quanto Branco: Estudos de Relações Raciais*. São Paulo: T.A. Queiroz.

O'Dougherty, M. 2002. *Consumption Intensified: The Politics of Middle Class Daily Life in Brazil*. Durham: Duke University Press.

Sansone, L. 2003. *Blackness Without Ethnicity: Constructing Race in Brazil*. New York: Palgrave Macmillan.

Schwarcz, L.M. 1993. *The Spectacle of the Races: Scientists, Institutions, and the Race Question in Brazil 1870–1930*. New York: Hill and Wong.

Schwartzman, L.F. 2007. Does Money Whiten? *American Sociological Review* 72: 940–963.

Sheriff, R.E. 2001. *Dreaming Equality: Colour, Race, and Racism in Urban Brazil*. New Brunswick: Rutgers University Press.

Skidmore, T.E. 1999. *Brazil: Five Centuries of Change*. Oxford: Oxford University Press.

Sutton, B. 2010. *Bodies in Crisis: Culture, Violence and Women's Resistance in Neoliberal Argentina*. New Brunswick: Rutgers University Press.

Telles, E.E. 2004. *Race in Another America: The Significance of Skin Color in Brazil*. Princeton: Princeton University Press.

Travae, M. White Is Right? General Maurão, An Indian and Candidate for Vice President of Brazil, Comments on How Handsome His Grandson Is and that He Is An Example of the 'Whitening of the Race.' *Black Brazil Today*, October 9, 2018. https://blackbraziltoday.com/white-is-right-general-mourao/. Accessed 4 Aug 2021.

Vargas, J.H.C. 2016. Black Disidentification: The 2013 Protests, Rolezinhos, and Racial Antagonism in Post-Lula Brazil. *Critical Sociology* 42 (4–5): 551–565.

Wade, P. 1997. *Race and Ethnicity in Latin America*. London: Pluto Press.

Wolf, N. 1991. *The Beauty Myth: How Images of Beauty Are Used Against Women*. New York: William Morrow and Company.

CHAPTER 8

Conclusion

The election of Jair Bolsonaro—who is known for his Far Right politics and defence of the former Brazilian military dictatorship—as President of Brazil in 2018 signalled for many commentators a radical shift in political discourse since democracy was restored in Brazil thirty years ago.[1] Promoting an ideology of social conservatism and nationalism, Bolsonaro's political rhetoric and views on women, race, immigration, abortion, and homosexuality (among other topics) have caused controversy and concern in Brazil and elsewhere (Butterworth 2018). According to Africana Studies scholar Keisha Khan Perry, the current political moment in Brazil should not be viewed as a radical departure from the past, however. Rather, it must be understood against the backdrop of a long history of white supremacist violence, anti-blackness, and the legacy of colonialism and slavery in Brazil (Perry 2019, 158). This history enacted repeatedly is directed at controlling, excluding, and removing the country's poor, Afro-Brazilian and indigenous people—a sentiment captured by Brazilian activist Abdias do Nascimento's quote in Chap. 3. Several contemporary scholars, such as Perry (2019), Smith (2016), and Vargas (2012), have convincingly argued that black, indigenous, and poor communities have long dealt with pervasive racialized and militarized violence, as well as

[1] Londoño, E. and Darlington, S. "Jair Bolsonaro wins Brazil's Presidency, in a Shift to the Far Right," *New York Times,* October 28, 2018. https://www.nytimes.com/2018/10/28/world/americas/jair-bolsonaro-brazil-election.html

D. J. Gordon, *Blackness and Social Mobility in Brazil,*
https://doi.org/10.1007/978-3-030-90765-5_8

everyday "structural violence," including segregated neighbourhoods in some areas, separate elevators for the public in apartment buildings, and austerity measures such as cuts to universities, public health care, education, housing, and cash transfer benefits.[2] Increased efforts under the present administration to repeal social gains made for blacks, women, and poor people under previous administrations (such as redistributive programmes and policies aimed at reducing inequality and poverty and affirmative action programmes) must be understood through this historical lens (Perry 2019, 159).

Although Bolsonaro has strongly denied being a racist, he has been associated with racist statements. In reference to *quilombolas* (the black descendants of rebel African slaves), he has said that, "they don't do anything. I don't think they're even good for procreation anymore" (Butterworth 2018). While he was a presidential candidate, he promised:

> You can be sure that if I get there [the presidency], there'll be no money for NGOs. If it's up to me, every citizen will have a gun at home. Not one centimetre will be demarcated for indigenous reserves and quilombolas.[3]

Despite this kind of disturbing rhetoric, Bolsonaro gained over half the country's votes in the 2018 national elections, attracting support from a variety of different groups such as Evangelicals, businesspeople, those disaffected with the Workers' Party, those who wanted tougher crime policies, as well as the farming lobby.[4] It is exactly this kind of political rhetoric and action that leads us to pose questions about racial discourses and how these shape societies as well as how to dismantle global manifestations of racism and racial discrimination that continue to have repercussions on the lives of African descendants.

[2] I take the term "structural violence" from Farmer (2003, 8), where he defined it as "a broad rubric that includes a host of offensives against human dignity: extreme and relative poverty, social inequalities ranging from racism to gender inequality, and the more spectacular forms of violence that are uncontestably human rights abuses, some of them punishment for efforts to escape structural violence."

[3] Ibid.

[4] Jair Bolsonaro won 55.1 per cent of the vote against Workers' Party candidate Fernando Haddad, who gained 44.9 per cent of the vote. See "Brazilian Election Results," Bloomberg, October 28, 2018. https://www.bloomberg.com/graphics/2018-brazil-election/. Also see Wilson, Katy "Jair Bolsonaro: Who Supports Brazil's New President?" BBC News, October 29, 2018. https://www.bbc.com/news/world-latin-america-45979682

The rise of emerging non-white elites in countries such as Brazil has brought new attention to their position within global and local hierarchies as well as renewed questions about the role of their middle classes in development. Emerging non-white elites pose a challenge to the long-standing, two class model of world society with white rich countries (the West) and poor, darker countries and regions (the rest). Now this divergent unipolar model of the world economy is shifting towards a multi-polar convergent version. I have focused in this book on the question of race, social mobility, and citizenship in a specific location, which should spark further discussion on the urgent questions of global anti-racist mobilization, struggles for social justice, and reparations to address long-standing structural inequalities in societies. It is highly debateable whether emerging non-white elites will play a revolutionary role in advancing democracy, racial equality, and development within their countries, or will simply act to conserve inequality in different forms. In South Africa and Caribbean countries, the black middle class belongs to the national majority, yet there is much debate about the nature of this mostly state-made elite. In South Africa, this group is referred to as "Black Diamonds," a term reflecting the lavish lifestyle and high-end consumerism of the politically empowered elite. Some (such as Mbeki 2009) claim that these elite are more alienated than ever before from the masses of poor, black people.

In Brazil, blacks are not the political or economic majority, but they have managed to carve out strategic positions within the state and in other areas of life. Meanwhile, the white elite have reorganized themselves, moving from a more rural base in land and industry, commerce, and retail business, while maintaining their hold on the economy, politics, and the media. Many of them have learnt to change with the changing times, aligning themselves with new political expressions of blackness that began to emerge in Salvador in recent decades. However, patron-client styles of relationship between the white elites and the majority black population reflects traditional forms of power. Yet specific avenues of mobility have been created around strategically placed black leaders, activists, and businesspeople, such as community projects that involve black middle-class people, for example. These, in turn, require specific responses—for example, emphasizing one's black identity is advantageous if one is aligned with projects sponsored by the Black Movement or culture industry in Salvador. Nevertheless, there is resistance to this in the very communities that they serve.

In this book, I have tried to show what kinds of people become involved in their communities and through what avenues. Also, I have tried to reveal the different strategies they use as they navigate between society, the state, and the market in order to improve their lives. Black Brazilians, however, may also negotiate Salvador's race and class hierarchies through less radical processes, such as through manipulating physical appearance and self-presentation. I have indicated the kinds of racial inequality they face in their everyday lives and how they deal will this by combining different social, cultural, political, economic, and aesthetic strategies. I have also identified some kinds of associational life that help to mitigate class and race differences, such as involvement in religious organizations and family networks.

Although my ethnography is limited in many ways to the specific context of Salvador, the material presented here throws light on the general issue of how black people experience local and international hierarchies of race, ethnicity, and culture. Indeed, Salvador's specific local system of classification carries within it echoes of a global order that has been shaped for centuries by slavery and colonial empire. What happens to black populations (including, perhaps especially, their new middle classes) in Africa and around the world poses a potentially revolutionary challenge to structures of inequality in the world at large and all of this takes place at a time of changing global relations and power dynamics in the century to come.

References

Bloomberg. 2018. Brazilian Election Results. October 28. https://www.bloomberg.com/graphics/2018-brazil-election/. Accessed 8 August 2021.

Butterworth, Benjamin. 2018. "Jair Bolsonaro: 17 Quotes that Explain the views of Brazil's Fascist President-Elect." The I, October 29, https://inews.co.uk/news/world/jair-bolsanaro-quotes-brazil-election-2018-result-president-elect-215221

Farmer, P. 2003. *Pathologies of power: Health, human rights, and the new war on the poor.* Berkeley: University of California Press.

Londoño, E. and S. Darlington, 2018. Jair Bolsonaro wins Brazil's Presidency, in a Shift to the Far Right. *New York Times,* October 28. https://www.nytimes.com/2018/10/28/world/americas/jair-bolsonaro-brazil-election.html. Accessed 8 Aug 2021.

Mbeki, M. 2009. *Architects of poverty: Why African capitalism needs changing.* Johannesburg: Picador Africa.

Perry, K.K.Y. 2019. The resurgent Far Right and the black feminist struggle for social democracy in Brazil. *American Anthropologist* 22 (1): 157–162.

Smith, C. 2016. *Afro-paradise: Blackness, violence, and performance in Brazil.* Urbana: University of Illinois Press.

Vargas, J.H.C. 2012. Gendered anti-blackness and the impossible Brazilian project: Emerging critical black Brazilian studies. *Cultural Dynamics* 24 (1): 3–11.
Wilson, K. 2018. Jair Bolsonaro: Who Supports Brazil's New President? BBC News, October 29. https://www.bbc.com/news/world-latin-america-45979682. Accessed 8 Aug 2021.

Appendix

Diasporic Positionalities

One of the concepts that can help us to critically examine racial negotiations across the African diaspora is "double consciousness," a term put forward by the American sociologist W.E.B. Du Bois. Double consciousness refers to the ways in which black people are forced to not only view themselves from their own perspective but to see themselves through the eyes of the outside (white) world. According to Du Bois, African Americans are always struggling to reconcile their "two warring souls," yet this could also be a source of tremendous insight—which he labels "second sight" (Du Bois 1990). This refers to the capacity of African Americans to see incomparably further into race relations when compared to white Americans and to speak about American society with heightened moral validity. Du Bois' notion of "double consciousness" and "second sight" applied not only to African Americans but to all people who are constructed outside of the dominant, white paradigm.

I hoped to use my positionality as a Caribbean scholar raised in a creole society to develop this "second sight." Indeed, Brazil and Jamaica have both had national myths of racial mixing and have been described as part of

D. J. Gordon, *Blackness and Social Mobility in Brazil*,
https://doi.org/10.1007/978-3-030-90765-5

a larger "Plantation-America culture sphere."[1] I thought that this similar context would provide me with unique opportunities for understanding racial negotiations in Brazil. While in some senses this was the case, in other cases this was not so. My subjectivity in the field was more like managing multiple warring identifications and different layers of ambiguity and power. I found it useful to read the writings of scholars who have critically examined the intersection of racialized, gendered, and national identities as they carried out research in Brazil (Caldwell 2007; Diversi and Moreira 2009; Hordge-Freeman 2015; Perry 2013; Twine 1998). Their reflections on the embodied reflexivity of lived experience resonated with me. How my body and other signifiers were given meaning and interpreted confirmed for me the centrality of race, gender, and body politics in Salvador. For example, on one occasion during fieldwork, I visited a Black Movement organization and was berated by a woman for wearing my hair blow-dried straight—to her this was a sign that I did not accept my blackness. She queried about my accent and I revealed that I was Jamaican. She looked even more surprised, exclaiming that I could not be an authentic Jamaican as I would have dreadlocks. Relating these experiences gives insight into the embodied and racialized experiences of diasporic research.

Being a "Betweener," Insider, and Outsider

Two Brazilian anthropologists, Diversi and Moreira (2009), have written a co-constructed narrative about their fieldwork encounters in Brazil which I find relevant to my experience in the field. They have described their engagement with marginalized populations in Brazil as white, male, Brazilian academics (albeit with different social backgrounds), using the concept of "betweenness" (Diversi and Moreira 2009, 13–27). They argue that this term has the potential to break down binary oppositions and ideological systems—such as an us/them, either/or view of the world—thereby creating possibilities for finding common ground and dialogue across differences. The concept of "betweenness" closely captures the sense of socially constructed, fluid spaces and identities that characterized my

[1] The "Plantation culture sphere" refers to the historical importance of the sugar plantation and slavery on the society and culture of the area, resulting in many common problems and similar contemporary traits throughout the region, such as hierarchical social relations and African retentions in cultural forms such as music, food, and religion (see Charles Wagley 1957, 5).

movements between Jamaica where I live, the United Kingdom where I was studying, and Brazil where I conducted doctoral research. The experience of occupying an in-between identity occurred on different levels. For example, while conducting fieldwork in Brazil, some persons saw me either as coming from a poor, black country or as privileged, especially because I attended a European university and was able to travel. As English was my native tongue, my status as a foreigner was often reinforced. I was also a single woman in Brazil without a family—a status that could be interpreted as being sexually available and potentially threatening to married women. It was interesting how this threat was mitigated by my sister's visit to Brazil in the latter months of my fieldwork. My Brazilian friends could then place me, locate me within a family—albeit far away in Jamaica.

My "betweener" status in Brazil was reflected in the reality that I never occupied any single identification uniquely. However, as a foreigner conducting fieldwork in Brazil, I had opportunities to cross the boundaries of a socially and spatially divided city. For example, through the graciousness of a friend, I lived in a luxury apartment in a wealthy neighbourhood for a few weeks. This provided me with a sense of the lifestyle of the wealthy, many of whom hired maids, gardeners, butlers, and nannies to organize their lives and had access to exclusive services, such as ocean piers and boats, restaurants, and clubs. After a few weeks, I moved in with a family in a working-class neighbourhood where I hoped to understand social mobility from below. I followed my growing network of Brazilian friends to Candomblé temples[2] and Evangelical churches in working-class districts, restaurants and bars in popular spots around the city centre, children's birthday parties in luxury apartment buildings, family vacations on the islands off the coast, and street parties and festivals, among other activities. Each of these contexts offered different spaces for socialization and required a different *habitus* and way of relating to persons within the specific situation. Although I was a "betweener," hovering in the uncertainty of shifting positions, at times I also actively guided these transitions, for example, by manipulating my personal style and appearance and by arranging house parties to which I invited my Brazilian friends. At times, they helped me to become less of a foreigner, such as encouraging me to buy Brazilian clothes.

[2] Candomblé temples, which are led by a priest or priestess, are often called "*terreiros*" or "*casas*" (houses). Temples are often the homes of Candomblé followers and are used for religious purposes. Candomblé is an Afro-American religious tradition, practiced mainly in Brazil, often combining elements from African religion and Catholicism.

However, I also had some unexpected experiences. For example, I lived for most of my fieldwork in a self-contained apartment in the house of an Italian Brazilian, centrally located near to a major transport hub and close to the homes of key respondents. I was invited to meals he prepared along with his white Brazilian friends. On one of these occasions, one of his friends seemed perfectly at ease telling me about "dirty black people who smell and do not bathe." To her, I was not *negra* (black) in the Brazilian sense. I was perhaps an honorary white, a foreigner living in the house of her friend. I learned not to react in shock or anger to similar such comments, but to make key inferences about racial discourse and white habitus in Brazil.

As in most research contexts, the dynamic between myself and my research participants influenced the information they chose to reveal, or not to reveal. For example, my skin colour in some contexts was associated with whiteness and privilege, affecting how people in turn described themselves or others to me. However, I also found that as a Jamaican, lines of communication quickly opened with most people. Jamaica holds a special, symbolic place in Salvador as a country associated with the articulation of a strong black identity and consciousness. Many people wear T-shirts with Jamaica printed on it, and identified strongly with Reggae music, especially that of the Marley family, Jimmy Cliff, Gregory Isaacs, and Burning Spear. During my fieldwork, Pelourinho—the historic city centre—had a Reggae bar, a Jamaican restaurant and shop, and a carnival *bloco* (band) called Olodum that identified with and used the colours of the Rastafari religion. I certainly was able to use this symbolic capital in a variety of social settings in Salvador. I also found that my position as an English speaker from a culture Brazilians considered to be *legal* (cool), often placed me among a privileged minority who travelled, studied, and worked in global settings.

Selection of Families and Individuals

I met the families that were involved in this research largely through snowballing techniques. Initially, I worked with faculty members and graduate students at the Federal University of Bahia who assisted me with meeting families and individuals who self-identified and were identified by others as middle class. In these initial months, my criteria for selection of the families and individuals in the research depended heavily on income, neighbourhood residence, and occupation (mostly civil servants and professionals). However, as the fieldwork progressed, I became more flexible with my criteria and decided to live for a few months in a low-income neighbourhood to see if I could understand mobility from below. I lived with a family in Liberdade and moved along kinship and friendship networks, eventually meeting individuals who were or became upwardly mobile in later years. After a few months, I moved to Campo Grande, which was more centrally located and close to a major transport centre. This change of residence opened opportunities to meet people of different social backgrounds through regular activities such as going to the gym, supermarket shopping, participation in dance or Capoeira classes, visits to theatres and museums, and attendance at festivals and events.

After about eight months of research, I was working with seven core families and their kinship/friendship networks whose members lived in different neighbourhoods. I met with them in various places to carry out interviews, including in their homes, at their work-places, in public areas like the mall or in a cafe, or at my home. I also met a key informant who gave me the names and numbers of individuals involved in the Black Movement whom he regarded as black middle class. I made some contacts here and successfully carried out some interviews with the members and leaders of Afro-Brazilian organizations.

Interviews

Moving to a new country and learning to speak a different language was an adjustment. My initial interviews tended to be more structured and formal; because of this, I devised questions along with my research assistant and tried them out on Brazilians I met to determine whether people could understand me and to see how they would respond. These pre-interviews are not included in this study. As I got to know people whom I wanted to include in the study, I discussed with them the research I was carrying out as well as any privacy concerns. Privacy was important because some of the questions I asked were sensitive and could potentially cause conflict if the topics were discussed openly. Only in exceptional or very obvious cases were the real names of the person used. Permissions were sought for recordings or photographs, as well as in cases where I wanted to raise a point mentioned in someone's interview with others.

I created a short interview schedule based on my previous experience, comprising basic questions about participants' occupational trajectory, income, and education. These structured interviews were also useful for contexts when I had to interview research participants in their offices when they had limited time. Eventually, I introduced longer, in-depth interviews, such as life histories and genealogies. These sessions were more free-flowing, conducted either with one or with more than one family member—frequently with the aid of memorabilia such as photographs, family albums, and objects or gifts from childhood. In many cases, these interviews involved more than one session and follow-up questions. Life histories and genealogies revealed rich and detailed information on the experiences, meanings, perspectives, feelings, emotions, and respondents' reactions to events.

I also utilized the reflexive dyadic interview, where I selected the questions and avoided interfering or changing the story that was told. I often double checked my information or shared transcriptions of interviews with respondents in a subsequent, private interview. This allowed for more accuracy in understanding, respectful dialogue, and self-determination by respondents. As Ellis (2004) has argued, these traditional interview situations are useful but not inherently superior to stories told in other situations. Finally, I utilized the "interactive interview," which has been described as a more self-conscious, collaborative process than reflexive, one-on-one interviews (Ellis 2004, 64). For example, sometimes I went to the homes of participants along with another researcher based in Salvador. We had small group discussions along with family members about topics such as their perceptions of social life in different neighbourhoods across the city, access to education and jobs, and views on race and racial discrimination. These were topics to which we could all relate. These types of interviews were followed up by a combination of informal discussions and one-on-one interviews.

Ethnography

This project is a qualitative and inductive study focused on racialized discourses and processes of social mobility and how research informants understand, negotiate, resist, or accommodate racialized hierarchies in Salvador. Stepping outside of the traditional idea of doing community-based research, my fieldwork was multi-sited and involved socializing with research participants as much as possible in different neighbourhoods, going with them to church, sports events, family parties, restaurants or bars, in addition to attending public events such as protests, university debates on racial quotas, festivals, shows, carnival events, and religious festivities. I adopted various roles in different settings, such as being a volunteer in a local church or helping to cook for family celebrations, or being an English instructor. I also lived with families for short periods in different stages of my fieldwork or went on family vacations, which allowed for detailed observations of everyday life in families. My ethnographic fieldwork was complemented by semi-structured and unstructured interviews of different kinds (detailed in the section on Interviews) as well as informal conversations.

My initial fieldnotes consisted of general comments and observations, such as making notes about household expenditures or writing down my impressions of social interactions and the sights, sounds, and smells of the city. At first, my fieldwork involved frantic searches to locate research participants, questioning busy academics, finding a research assistant, getting used to moving around the city and taking public transport, and considerable anxiety about not finding enough people for my study. However, I found most people to be helpful along the journey. A few of them refused to participate in the study or preferred just to be friends. Eventually, I settled on some core families who I visited regularly, although I got to know some better than others. It required a lot of physical energy to travel to visit with research participants in different neighbourhoods across the city and to socialize in different contexts. To create a better balance, my visitations became more spaced out than they were in the beginning and I gave myself more time between interviews to process the considerable amounts of data I had gathered. Other settings, such as Black Movement organizations, required formal appointments and interviews and in one instance, involved a focus group discussion with students attending an educational programme geared towards getting them successfully placed in public universities. In this case, I sought the services of a research assistant to help with coordinating and facilitating the focus group while I planned the interview questions and taped the proceedings.

Data Analysis

As a researcher involved in observing how structural inequalities affected everyday social interactions, I paid close attention to how informants made meaning of racialized contexts and how they translated those meanings into practices. Therefore, this study emphasized detailed ethnographic observations and what informants said in their interviews, in order to understand informants' cognitive maps of everyday life in the city. This involved an understanding of symbolic interaction in particular settings as well as wider social change over time. Therefore, gathering data from household and census surveys, studies of race relations in Salvador, archival research of Black Movement organizations, and information about the wider society from films, plays, theatre productions, and public events was an important part of the study.

In addition to recording and transcribing my interviews, I kept fieldnotes and a field dairy, which was a considerable amount of data to go through (there is still raw data that I have not touched!). I found that coding was an essential method used to organize key concepts and themes across my transcribed interviews and fieldnotes. I used colour coding and highlighters as well as a three-phase process involving open, axial, and selective coding. Key concepts consisted of single words or labels (e.g., "occupation" or "education") as well as generalized themes that emerged from reading my notes line by line, such as "attitudes towards hair." In axial coding, I emphasized how situational factors affected whether and when persons engaged in certain discourses or practices related to racialized differences—so this would include notes from my diary about people's responses, body language, or exploring what criteria determined how a person's actions were judged by others (e.g., if a family member was approached about their obligation to fund a family event, I would make a note of how others reacted and assessed this family member if they avoided or outright rejected this obligation). Selective coding was used to relate practices and discourses recorded in my notes to concepts central to my theoretical or conceptual framework. For example, this could include the behaviours and verbal messages involving phenotypic differentiation and relating it to racial formation theories or critical race or feminist perspectives. I also created an overall codebook on my computer that included the different codes and sub-codes and the page numbers in my fieldnotes where the raw data could be found. These coding processes helped me to develop, refine, and revise my themes, in addition to reading and re-reading my notes and interviews several times. Finally, what was very helpful to me was that I had to write quarterly fieldwork reports for my supervisors for which I received insightful feedback. Not only did this force me to reflect and be engaged with data analysis throughout the field research, it was useful for discovering new directions I needed to take (or to cease going down unproductive pathways). When I returned from fieldwork to begin writing up my doctoral thesis, these fieldwork reports and my supervisors' feedback were remarkably useful for their analytical insights.

References

Caldwell, K. 2007. Negras in Brazil: Re-envioning Black Women, Citizenship, and the politics of Identity. New Brunswick: Rutgers University Press.

Diversi, M. and C. Moreira. 2009. Betweener Talk: Decolonizing Knowledge Production, Pedagogy, and Praxis. New York: Routledge.

Du Bois, W.E.B. (1903) 1990. The souls of Black Folk. New York: Vintage Books.

Ellis, C. 2004. The Ethnographic Eye: A Methodological Novel about Autoethnography. New York and Oxford: Altamira Press.

Hordge-Freeman, E. 2015. The Color of Love: Racial Features, Stigma and Socialization in Black Brazilian Families. Austin: University of Texas Press.

Perry, Keisha-Khan Y. 2013. Black Women Against the Land Grab: The Fight For Racial Justice in Brazil. Minneapolis: University of Minneapolis Press.

Twine, France W. 1998. Racism in a Racial Democracy. New Brunswick: Rutgers University Press.

Wagley, C. 1957. "Plantation America: A Culture Sphere." In Vera Rubin (ed), Caribbean Studies: A Symposium. Seattle and London: University of Washington Press, 3–13.

Index[1]

[1] Note: Page numbers followed by 'n' refer to notes.

D. J. Gordon, *Blackness and Social Mobility in Brazil*,
https://doi.org/10.1007/978-3-030-90765-5

N

O

P

The manufacturer's authorised representative in the EU is Springer Nature Customer Service Centre GmbH, Europaplatz 3, 69115 Heidelberg, Germany. If you have any concerns regarding our products, please contact ProductSafety@springernature.com

Printed and bound by CPI Group (UK) Ltd, Croydon, CR0 4YY
22/09/2026
02221067-0009